DEDICATION

We dedicate this book to our families, who put up with being our test subjects and with the brooding writers that we become as deadlines approach. We love you all, and we thank you for your patience with us!

THE SASSAFRAS SCIENCE ADVENTURES

VOLUME 2: ANATOMY

JOHNNY CONGO & PAIGE HUDSON

THE SASSAFRAS SCIENCE ADVENTURES
VOLUME 2: ANATOMY

Second Printing 2015
First Edition 2013
Copyright @ Elemental Science, Inc.
Email: info@elementalscience.com

ISBN: 978-1-935614-24-1
Cover Design by Paige Hudson & Eunike Nugroho
Illustrations by Eunike Nugroho (be.net/inikeke)

Printed In USA For World Wide Distribution

For more copies write to :

Elemental Science
610 N. Main St., Suite 207
Blacksburg, VA 24060
info@elementalscience.com

Table of Contents

MAKE THE MOST OF YOUR JOURNEY WITH THE SASSAFRAS TWINS!

Add our activity guide, logbook, or lapbooking guide to create a full science curriculum for your students!

The Sassafras Guide to Anatomy includes chapter summaries and an array of options that coordinate with the individual chapters of this novel. This guide provides ideas for experiments, notebooking, vocabulary, memory work, and additional activities to enhance what your students are learning about the human body!

The Official Sassafras SCIDAT Logbook: Anatomy Edition partners with the activity guide to help your student document their journey throughout this novel. It includes their own SCIDAT log pages as well as body system overview sheets and an anatomy glossary.

Lapbooking through Anatomy with the Sassafras Twins provides a gentle option for enhancing what your students are learning about the human body through this novel. It contains a reading plan, templates and pictures to create a beautiful lapbook on anatomy, vocabulary, and coordinated scientific demonstrations!

VISIT SASSAFRASSCIENCE.COM TO LEARN MORE!

THE SASSAFRAS SCIENCE ADVENTURES

THE SASSAFRAS GUIDE TO THE CHARACTERS

THROUGHOUT THE BOOK:

(*Note* – These characters also appeared in *The Sassafras Science Adventures Volume 1: Zoology*.)

* **Blaine Sassafras** – The boy Sassafras twin, also known as Train. He started the summer hating science, but is now starting to change his mind.
* **Tracey Sassafras** – The girl Sassafras twin, also known as Blaisey. She started the summer hating science, but is now starting to change her mind.
* **Uncle Cecil** – The Sassafras twins' crazy, forgetful, and messy uncle. He is the scientist behind the invisible zip lines.
* **President Lincoln** – Uncle Cecil's lab assistant, who also happens to be a prairie dog. He is also the co-inventor of the zip lines.
* **The Man with No Eyebrows** – He has no eyebrows and an extreme dislike for Uncle Cecil. Not only is he spying on the red-haired scientist, but he is also trying to sabotage the twins at every stop.

ADDIS ABABA, ETHIOPIA (CHAPTERS 2-3):

* **Larry 'Snowflake' Maru** – (mah-roo) The local expert for the Ethiopian leg of the twins' anatomy adventures. He dabbles in archaeology and is very knowledgeable about the skeletal system.
* **Raz** – He is Snowflake's best friend and owner of Raz's Pawn & Antiques shop.
* **Retta** – She is Raz's wife and co-owner of the Raz & Retta's coffee shop.
* **Brother Eskinder** – He is the Ethiopian custodial priest of the Holy Trinity Cathedral.

SYDNEY, AUSTRALIA (CHAPTERS 4-5):

* **Julie Ette** – The local expert for the Australian leg of the twins'

anatomy adventures. She is an amazing singer participating in the singing portion of the Take Your Breath Away competition.

✶ **Suzy McSnazz** – She is one of the finalists in the singing portion of the Take Your Breath Away competition.

✶ **Colorado Quadruplets (aka Ladder Smash)** – Denver, Dexter, Denise, and Dolores make up this act, which is one of the finalists in the dancing portion of the Take Your Breath Away competition.

✶ **Trudy Stiles** – She is one of the finalists in the dancing portion of the Take Your Breath Away competition.

✶ **Flip Pippen** – He and his puppet, Zippy, are finalists in the magic portion of the Take Your Breath Away competition.

✶ **The Dark Cape** – He is one of the finalists in the magic portion of the Take Your Breath Away competition.

✶ **Bob Squats** – He is one of the finalists in the miscellaneous portion of the Take Your Breath Away competition.

✶ **Cletus Magnolia** – He is one of the finalists in the miscellaneous portion of the Take Your Breath Away competition.

✶ **Dean Bean, Jr.** – The announcer for the Take Your Breath Away competition.

✶ **Victoria Valencia, Bobbie Mega, Miles Dockerty** – The judges for the Take Your Breath Away competition.

VENICE, ITALY (CHAPTERS 6-7):

✶ **Vittorio Benaneli** – (vee-tor-ee-o ben-ah-neh-lee) The local expert for the Italian leg of the twins' anatomy adventures. He is an expert chef and owner of Benaneli's restaurant.

✶ **Giovanni** – (gee-o-van-ee) He is Vittorio's nephew and dishwasher at Benaneli's.

✶ **Salvatore & Bruno** – They

are Vittorio's competition seeking to put him out of business by learning his secret ingredient.

* **Mama Benaneli** – She is Vittorio's mother, Giovanni's grandmother, and keeper of the secret ingredient.

BEIJING, CHINA (CHAPTERS 8-9):

* **Coach Boxton** – The local expert for the Chinese leg of the twins' anatomy adventures. He is also the American coach of China's Olympic decathlete team.
* **Jek** – He is Coach Boxton's assistant and translator.
* **Dr. Veeginburger** – He is a pharmaceutical salesman who tries to sell Coach Boxton illegal supplements.
* **Itsy** – He is Dr. Veeginburger's muscle and proof that the supplements work.
* **The Ancient Calligrapher** – He is a legendary old man who secretly lives in the Great Wall of China. He is said to possess great wisdom but will only answer one question.

LUBBOCK, TEXAS (CHAPTERS 10-11)

* **Burly Scav** – The local expert for the Texan leg of the twins' anatomy adventures. He is a local trash man and inventor of the Smart Dump.
* **Trevor Scav** – He is the teenage son of Burly Scav.
* **Kimlee Broadstine** – She is an investor from the Broadstine Investment Group and the daughter of Broderick Broadstine. She unwillingly reviews the Smart Dump project for potential investors.
* **Smirk & Chili** – They are two of the meanest, nastiest, but coolest kids from Trevor's school.

BANGKOK, THAILAND (CHAPTERS 12-13)

* **Dr. Olivia Apple** – The local expert for the Thai leg of the twins' anatomy adventures. She is the director of the woman's center in a local hospital.
* **Nart** – (nah-rr-t) He is the accident prone Thai man who is always hanging around Dr. Apple's hospital.

* **Chanarong** – (ch-ahn-rr-ong) He is an officer in the Thai military and friend of Dr. Apple.
* **Lawana** – (lah-wah-nah) She is the pregnant friend of Dr. Apple. She has been kidnapped. Her husband, Mongkut, died in a construction accident.
* **Kingman Nawarak** – (nah-wah-rr-k) He is an evil man who tries to control Bangkok through cheating and kidnapping.

ALASKA (CHAPTERS 14-15)

(**Note** – These characters also appeared in *The Sassafras Science Adventures Volume 1: Zoology*.)

* **Summer Beach** – The local expert for the Alaskan leg of the twins' anatomy adventures. She was also a former classmate of Uncle Cecil's and has a bit of a crush on him.
* **Ulysses S. Grant** – Summer Beach's lab assistant, who also happens to be an arctic ground squirrel and inventor extraordinaire.

DUBAI, UAE (CHAPTERS 16 & 17)

* **Sylvester "Doc" Hibbel** – The local expert for the Dubai leg of the twins' anatomy adventures. He is a traveling salesman and inventor of several medicinal elixirs.
* **Sheikh Rehan** – He is a billionaire oil baron and a big fan of cowboy culture as well as horse racing.
* **Arnie Derbinhoogan** – He and his horse, Horsinhoogan, are contestants in The Wind Tower 100 race.
* **Najib** – He and his horse, Yazer, are contestants in The Wind Tower 100 race.
* **Itja** – (eet-jah) The scoundrel leader of a group of bandits, known as the Kekeway (Kee-kee-way). He also appeared in *The Sassafras Science Adventures Volume 1: Zoology*.

VOLUME 2

ANATOMY

CHAPTER 1: ADVENTURES IN ANATOMY

Breakfast at Cecil's

The smell of sizzling bacon and mothballs reached the noses of sleeping twelve-year-old twins, Blaine and Tracey Sassafras. Blaine, the older of the twins by five minutes and fourteen seconds, began blinking his sleepy eyes open.

"Wait a second. That doesn't make any sense. Why would those smells of mothballs and bacon ever be mixed together? Where am I?" he said to himself. Blaine sat up and looked over at his sister who was waking from her own sleep.

"Tracey," croaked Blaine, a bit alarmed. "Where are we? This isn't our house!"

Tracey sat up, groggy-eyed, and looked around the unkempt room before she muttered, "Of course it's not our house, Blaine. Don't you remember? We are at crazy Uncle Cecil's house."

Blaine's mouth dropped open almost as wide as his eyes, "You mean that wasn't a dream?"

"No, it wasn't a dream," Tracey exclaimed, as she attempted to use her fingers to comb down her messy hair.

"So the invisible zip lines, all the animals, the Man with No Eyebrows, that was all . . . real?" Blaine mumbled in disbelief.

Both twins sat silent in their beds for a moment, thinking about all that had transpired so far this summer. Their parents had sent them to their Uncle Cecil's house to brush up on their scientific knowledge because of a failing grade in the subject. They had arrived at 1104 North Pecan Street expecting the worst, but it had turned out to be the beginning of the greatest adventure of their lives. Their well-known scientist uncle and his pet prairie dog, President

Lincoln, had invented invisible zip lines that had the ability to take them anywhere on the globe at the speed of light. All that was needed was a harness and a special three-ringed carabiner. One ring of the carabiner was for longitude coordinates, one ring was for latitude coordinates, and the third ring locked the carabiner closed. Once the carabiner was correctly calibrated to the right coordinates and snapped shut, it would automatically hook onto the correct invisible zip line. Then within a matter of seconds, whoosh, off you went on a sonic-speed journey to the desired location.

Over the course of the last week or so, the Sassafras twins had used these invisible zip lines to travel to Kenya, Egypt, Canada, Peru, Australia, China, Alaska, and the Southern Atlantic Ocean. They had encountered all kinds of amazing animals on their journeys and had met really incredible local animal experts at each location. They had used an application called SCIDAT (short for 'scientific data') on the smartphones that their uncle had given them to text in and store loads and loads of exciting and relevant information on the animals they encountered. Blaine and Tracey had survived some pretty amazing and perilous adventures on their travels and had made it back to their Uncle Cecil's house a whole lot smarter. The twins had even survived the Man with No Eyebrows's attempts to sabotage and stop them. He was a strange man who had somehow shown up at several of their locations around the globe. He seemed to be trying to destroy them. Neither Blaine nor Tracey knew why, but at least he hadn't succeeded.

When the twins had arrived back at their uncle's house after recording data on the giant squid, they had assumed they were finished with globe zipping for the summer and maybe even forever, but they had been wrong about that. Uncle Cecil had forgotten to tell them that their study on zoology was just the first of many science subjects they would be covering over the course of the summer. Now they had so much more to look forward to, more adventure to be had, and more science to learn. The twins were hungry for more.

They would be studying anatomy next, and if it was even half as cool as zoology had been, then Blaine and Tracey knew that they were in for a wild ride. Their stream of flowing memories was cut off by yet another strong smell. The aroma of oranges was now added to the mix. Blaine asked his sister, "Do you smell that?"

"What?" Tracey questioned. "The bacon, the mothballs, or the fresh squeezed orange juice?"

"So you smell all of that too?" Blaine confirmed, hopping out of bed. "I know the mothball smell is coming from this room Uncle Cecil put us in. Really, all the rooms upstairs reek of mothballs. It's gross! It smells like a grandma's sock drawer, but what about the bacon and the orange juice? Do you think Uncle Cecil is making us breakfast?"

"There's only one way to find out," Tracey responded, getting out of her bed. Both twins slipped on their shoes and prepared to head downstairs.

Their Uncle Cecil's house had obviously been decorated decades before the twins were born. It not only smelled like a grandma's house but it looked like one too, at least on the second floor where their room was located. Cecil spent most of his time down in the basement working on science projects and inventions, so he was rarely upstairs. He had told the twins the night before that most of the upstairs rooms were just full of junk. Their room, evidently, was the only room that had enough space to sleep in. The twins assumed he saved so much stuff just in case he could use some of it on one of his projects.

Blaine and Tracey made their way to the stairs, which were half-covered by stacks and stacks of cardboard, old newspapers, and books. They started going down toward the kitchen. The mothball smell lessened and both the bacon and orange juice smells intensified with every step down.

Blaine was the first to burst through the swinging kitchen door, with Tracey right behind him. They were shocked at what they saw. The whole kitchen had been converted into some kind of … breakfast-making machine. The kitchen had not looked like this the night before. There was bacon being fried by some sort of mechanical arm. There was a big cylindrical blender with a literal tornado of juice swirling inside. There were many mechanical moving parts all around, opening drawers, grabbing utensils, mixing ingredients, stocking and un-stocking the refrigerator. And there, standing on the countertop in the middle of it all, with a smile on his face and an apron tied around his waist, was none other than President Lincoln, the prairie dog.

Then, as the twins looked around in disbelief, several small doors opened up high on the far wall and released four eggs. The eggs rolled out the doors and landed on a moving track. The track carried the eggs safely around a series of curves and then dropped them, without breaking, through alternating sections of a breakaway bridge. At the bottom of the bridge, the eggs were safely caught by robotic fingers. The fingers did some fancy spinning and flipping of the eggs before cracking each egg and letting the contents ooze down into waiting skillets. The eggs fried nicely in a matter of seconds. The spring-loaded skillets then flipped the eggs through the air, landing them perfectly on waiting plates that were already on the kitchen table. Immediately, the mechanical arm swung around and added hot bacon to the plates. Then, the lid of the tornado blender popped off and a mechanical hand grabbed the container. It swung the juice around and poured it into the waiting glasses on the conveyor belt.

Tracey looked at Blaine with an impressed look on her face. "Well, I guess President Lincoln really is an inventor, just like Uncle Cecil said."

"Yeah, just look at this breakfast," Blaine said, enthusiastically. "Sizzling bacon, fresh-squeezed orange juice, cereal, and fried eggs! This is fantastic!"

President Lincoln jumped down off the countertop and joined the twins at the table as they prepared to eat breakfast. Then, like a ball of crazy red-headed energy, Blaine and Tracey's Uncle Cecil came bursting through the swinging kitchen door.

"Train! Blaisey! Tipper Topper of the morning to you! You two are looking bright-eyed and bushy-tailed on this fine day. Whoa! Linc dog, what have you done to the kitchen?" Cecil proclaimed to his furry friend. "It looks…great! I like it!"

Cecil swooshed around the table in his white lab coat and bunny rabbit slippers and sat down in front of the last plate on the table. He immediately began downing the breakfast, but that did not stop him from talking. Not at all.

"I am still beside myself with joy at how well you two did on all the zoology assignments from this summer's zip lining adventures. Oh, zippy zip zipparoo. The zip lines have worked great, wonderfully, fantabulously! Have they not? And there's more where that came from!"

The energy and excitement that their uncle possessed never

seemed to lose its luster, but he was especially hyper this morning. The twins wondered if he had already had his morning cup of coffee, or rather his morning pot of coffee! Or maybe he was just excited about the start of a new subject of science. Whatever the reason, he was like a regular tornado of happiness this morning.

"The two of you," Cecil announced, while stuffing more bacon into his mouth, "got an A+++ on zoology. Next up— anatomy! I know that you guys are going to do superiffic on this subject too. Well, my wonderful niece and nephew, are you excited? Are you ready to continue your zip line science adventure?"

Blaine and Tracey both nodded in excited confirmation.

"Well, then," Cecil declared as he shot up from the kitchen table; amazingly, he was already finished with his breakfast. "You two finish eating here, and then meet President Lincoln and myself down in the basement for a presentation that our genius prairie dog has prepared for you."

Cecil exited through the kitchen door as fast as he had come in. President Lincoln left the table as well and darted through a hole in the wall that led to a network of tunnels the prairie dog used to travel around the house. This left Blaine and Tracey alone at the kitchen table. Blaine took a swig of orange juice and looked over at Tracey.

"Well, sis," he remarked. "Are you really ready for more? I know that there is more adventure waiting on the other side of those lines for us, but there is also more science. What if we don't enjoy anatomy like we enjoyed zoology? What if anatomy is boring?"

Tracey took a bite of egg and thought for a moment before she replied, "Do you really think it's going to be boring, Blaine? Not one thing has been boring since we got off that bus and arrived at Uncle Cecil's house. I think anatomy is going to be just as fun as zoology was."

Blaine smiled, "Me too. I was just checking to see if we were still on the same page."

His smile turned into a smirk as Blaine gave Tracey a decent punch in the shoulder, hopped out of his chair, and then shouted, "Beat you down to the basement!"

Tracey, unable to turn down a challenge from her twin brother, hopped quickly out of her chair and ran toward the basement, hot on Blaine's heels.

It had been fairly easy to hide all the cameras in Cecil Sassafras's house. Most of them were also equipped with speakers so he could hear what was going on as well as see. The hardest place to hide the video cameras had been in the basement. Granted, it was so messy and unorganized down there that the loopy, red-headed scientist would never notice a few small hidden lenses here and there, but it had been difficult to find a time that Cecil wasn't in the basement. The man practically lived down there. In addition to spending hours and hours on various scientific projects, he also often ate and slept down there, and when Cecil wasn't down in the basement, that pesky prairie dog was. Even so, he had successfully done it.

Now, from the basement of his house at 1108 North Pecan Street, just two doors down from the Sassafras residence, he was able to watch every move that Cecil and his niece and nephew made. He had noticed that the children were referring to him as "The Man with No Eyebrows." A fitting title indeed, though somewhat offensive. Even though the kids called him that, Cecil had not yet seemed to figure out who he was. That was just like Cecil Sassafras—head always in the scientific clouds, oblivious to human relationships. If somebody loved Cecil, like that silly Summer Beach did, Cecil would miss all the obvious signs. And if somebody hated Cecil, like he did, Cecil would miss that too. But what Cecil couldn't possibly miss was the revenge that was going to be exacted on him by the "Man with No Eyebrows."

The villain chuckled to himself at the thought of this sweet vengeance, as he even now used the hidden cameras to watch those twins rumble down into the basement joining Cecil and that prairie dog. He assumed that the children were about to zip off to their next location, and when they did, he would be there. He would use the cameras to see the longitude and latitude coordinates listed in the LINLOC application on the kids' phones. Then he would use his own three-ringed carabiner and invisible zip lines to travel to the exact same location as the children. He would sabotage Cecil's niece and nephew's science learning. The man knew that this whole global zip lining, science-face-to-face thing was Cecil's dearest and most esteemed project ever. So he would take what was dearest to Cecil and destroy it.

Over the past week, he had traveled to different locations, and he'd tried to thwart the children, but they had proven to be very hard to stop, indeed. They were determined and smart, too, but he knew that determination could only last so long. He would think of bigger and better ways to stop them, and he would be relentless in his pursuit of revenge. Eventually, he would stop those Sassafras twins.

Socrates and Aristotle

President Lincoln jostled around with the basement computer's mouse, which in turn illuminated the tracking screen. The tracking screen was a map of the world that Uncle Cecil used to observe Blaine and Tracey's progress through their global locations. Two green dots represented the twins and those dots moved with them as they moved from location to location. Though they were traveling all over the planet with no adults, they felt pretty safe. They knew that Cecil was tracking their movement down here in the basement, and that they could use their smartphones to call him anytime they were in trouble. Besides, they were enjoying the adventure.

The only thing that worried them just a bit was that sneaky Man with No Eyebrows. They never knew when he was going to show up and try to sabotage them. They had bested him before and were pretty sure they could best him again if he ever popped up.

With another tap of the mouse from President Lincoln, the screen changed from the map of the world to the document page. This was where Cecil could see their SCIDAT information as the twins texted it in and also where he could view the pictures that they took. However, right now, the document page displayed a picture of the prairie dog with brightly colored text that read, "President Lincoln's ever so Brief Presentation on Zoology."

Blaine and Tracey still weren't sure what to think about the lovable President Lincoln. Was he really an inventor? Could he even truly communicate? They had seen what Lincoln had done to the kitchen this morning, and here he was, giving an electronic presentation. Maybe the prairie dog wasn't everything Uncle Cecil said he was, but at the very least he was the smartest animal they had ever known. Cecil read the text aloud as President Lincoln used the mouse to click through the pages of his presentation.

"The five major divisions of the animal kingdom," started Cecil, "mammals...."

As their uncle said, "mammals," a picture they had taken of an elephant in Africa came up on the screen. The twins looked at each other in surprised joy. President Lincoln had put together a presentation using the pictures they had taken. How cool was that!!

"Birds..." Cecil continued next as a picture of an Alaskan Snow Goose in flight came up on the screen.

"Reptiles..." Now a picture of a cobra from their would-be tomb in Egypt, appeared. The twins shuddered at the thought of how close they had come to being goners.

"Amphibians..." Cecil stated as the picture of the tiny poison dart frog they had seen on the bromeliad plant in Peru.

"And last but not least—invertebrates." The picture here was of a spider from the barn in Canada.

"Next, let us look at the three different types of animal diets. First, we have carnivores, which means meat eaters." As Uncle Cecil read this text, a picture of a racing cheetah appeared. The twins recalled how glad they had been when the cheetah had chosen a wildebeest for lunch instead of them.

"Then there are omnivores," Cecil said, "which means an animal that eats meat and plants." Now the twins saw a picture they had taken of a spiny-tailed lizard as it was warding off a sand cat.

"And thirdly, there are herbivores. Herbivores are animals that eat plants exclusively." The picture was one that Blaine had taken of a koala while in the Brown Mountain Forest of Australia.

The last page of President Lincoln's presentation came up. Cecil read its contents. "The Latin word for 'human' is '*Homo sapiens.*' *Homo sapiens* are mammals and omnivores."

Uncle Cecil smiled and clasped his hands together. "That concludes our furry little friend's presentation on zoology. It was a splenderiffic review, with some pretty amazing pictures, I might add."

He reached over and gave President Lincoln a high five. "Let us now move on to our next subject at hand—anatomy."

Blaine and Tracey felt the excitement building over this new subject. It was hard to believe they had ever disliked science.

"Before we start," Cecil declared, "I have some skeletons I need to get out of my closet." Their uncle walked over to a nearby closet door and opened it up. He pulled out two plastic skeletons.

"Meet Socrates and Aristotle," Cecil proclaimed, with a big grin.

"You named your plastic skeletons?" Tracey asked.

"Yessiree! I have had these guys longer than the two of you have been alive."

"Why is one missing a head?" Blaine asked.

"Well, I'll be a green persimmon," Cecil remarked, as he looked around. "Socrates, what did I do with your head? Oh yes, I lined it up on the table last night to represent anatomy." Cecil grabbed the skull off the table and attached it back onto Socrates's body of bones.

"OK, Sassafras twins, why don't you two pull out your smartphones and open up the LINLOC application to see where you are going first on this anatomy leg of your summer science adventures?"

Blaine and Tracey happily obliged by grabbing their backpacks, which they had left in the basement the night before.

Then, they pulled out their smartphones and used the touch screens to find the LINLOC application. This app was short for "Line Locations" and it gave them their intended locations, including the longitude and latitude coordinates, the name of their local expert, and the scientific topics that they needed to gather data on. Tracey read aloud with excitement what was listed on LINLOC right now.

"We are going to Ethiopia: Longitude 38° 42' E, Latitude 09° 02' N! Our local expert's name is Larry 'Snowflake' Maru, and we will be gathering information on skulls, backbones, bones, and joints."

"Ethiopia? Cool! That sounds like a good start to a new subject!" Blaine exclaimed.

"Indeed it does!" Cecil bubbled over, exuberantly.

He grabbed Socrates, the plastic skeleton, and started dancing around the basement. "So, you will start with the skeletal system in Ethiopia and progress through different locations, really fleshing out the *homo sapiens'* anatomy! Oh, Honky Tonky Bonk-O-Bonky! I am so excited for you two!"

Cecil stopped dancing with Socrates when the skeleton's head fell off again, but he continued with some more instructions. "Don't forget that the invisible zip lines are to be kept a secret. They are designed to land you as close to your local experts as possible without being detected. Also, I'm sorry to say, President Lincoln and I have not yet managed to fix the glitch."

"You mean the glitch that won't let us progress through the LINLOC locations if we don't enter the proper SCIDAT data?" Tracey asked.

LINLOC SCIDAT

LOCATION: Ethiopia
CONTACT: Larry "Snowflake" Maru
LATITUDE: LONGITUDE:
09°02'N   38°42'E

ETHIOPIA

INFORMATION NEEDED ON:
Skull, Backbone, Bones, and Joints

"Or the one that sent us to separate places?" Blaine added, remembering how he and Tracey had zipped off to two separate locations, when the Man with No Eyebrows had jumbled up their SCIDAT info.

"Persactly, that glitch," Cecil confirmed. "But on the brighter side of things, we did get a new application finished for you, and we will upload that to your phones right now!"

President Lincoln, who was still over by the computer, pushed a button and the twins saw that immediately the new app was being wirelessly uploaded to their phones.

"This is what I like to call the archive application," Cecil informed them, as he attempted to re-attach Socrates's head again. "In zoology, the two of you were able to take pictures of everything, but in anatomy and beyond that, this may not be possible. So this new application will give you the continued capability of sending in pictures with your SCIDAT information. Of course, you can still take pictures with your phones' cameras and send those in like before. However, when that is not possible, you can scan through the images in the archive app, select the one that is appropriate, and send that in with your data."

"Very cool!" Tracey nodded in understanding.

As the new app finished uploading, Cecil finally managed to click Socrates's skull back into place. He then clapped his hands once and then started wiggling his fingers in excitement.

"You have your carabiners. You have your harnesses. You have your backpacks and helmets. You have your smartphones, complete with the new application. Train and Blaisey! It is time to zip!"

"Longitude: 38°42'E, Latitude: 09°02'N," he wickedly whispered to himself, as he turned the rings of his own carabiner to the coordinates he had clearly seen on the children's phones through the lenses of his hidden cameras.

"OK, Sassafras twins," the Man with No Eyebrows rasped, with an edge to his voice, "I will see you in Ethiopia."

CHAPTER 2: ETHIOPIA, HERE WE COME!

The Surprising Skull

There was nothing like it! The feeling of zipping through places at the speed of light was amazing. Blaine and Tracey Sassafras were now soaring through rushing brightness, on invisible zip lines, headed toward the country of Ethiopia. Only one day ago, they had thought their zip lining adventures were over, but here they were now, gliding across the lines again and enjoying every second of it.

Their travel came to a jerking stop, their carabiners automatically unclipped from the lines, and their bodies slumped down, strength gone and eyes blinded. This was a feeling they had become accustomed to. It was how the landings always happened. The lack of strength and sight was only temporary and they knew both would quickly return. Meanwhile, the twins both felt around with their hands. Apparently, they had landed on some kind of sagging tarp, which seemed to be littered with quite a bit of junk. The Sassafrases could also hear the loud sound of a crowd of people. Where on earth had they landed?

As sight slowly returned, Blaine and Tracey saw they had indeed landed on a canvas tarp. It looked to be an awning, and it definitely wasn't the only awning in the area. There was what looked like an endless stretch of different colored awnings covering what looked like a very busy street, and it felt like a hot day.

The twins looked around on their canvas awning and saw lots of trash surrounding them. They were just under a window. It looked like people had been throwing trash out of their windows above, and it had landed on this awning. There were apple cores, old shoes, plastic bags, candy wrappers, popsicle sticks, and more. Only one thing really stood out from the rest, and that was an old tattered backpack. With his strength and sight now returned, Blaine reached over and grabbed the old backpack.

"Gross, Blaine," Tracey exclaimed, with a scrunched up nose. "What are you doing? Don't sort through that trash."

"It's not trash," Blaine responded, holding up the old backpack. "Look at this thing. It's cool. It's like some kind of old army bag or something."

The backpack was littered with holes, and the front of it was almost completely torn off, revealing the backpack's insides.

"Wow," Blaine said. "There is some kind of writing and some weird lines stitched on the inside of the bag, but it's not English. I can't read it."

"Blaine," Tracey sighed, getting annoyed at her brother. "Who cares about that old bag? I think we have landed in the middle of some kind of huge outdoor market. Let's just figure out a way to get off of this awning and start looking for our local expert. I feel like I'm trapped in a hammock full of trash!"

As if in answer to Tracey's declaration, the twins heard a loud tearing noise as the canvas awning began to rip.

"Oh, no," the twins shouted in unison.

All at once, the tear became a hole big enough for them to fall through, and the Sassafrases felt themselves being taken down by gravity. The twins tumbled down onto a stack of hand-woven baskets. Blaine landed on his stomach and Tracey landed on her side, both hitting hard enough to knock the air out of their lungs. The trash that had been on top of the awning rained down on them and on the baskets that they had landed in. Blaine looked up to see a very surprised woman looking at him and Tracey with wide eyes. He figured she must be a vendor in what was definitely a market, and these must be her baskets that were partially smashed and covered in garbage.

"Sorry," Blaine uttered, smiling sheepishly.

The woman raised her hands to the side of her head and looked like she was trying to shout, but was so shocked that nothing was coming out. Two guards had seen Blaine and Tracey fall through the awning and land on the baskets. They were now looking at the confused vendor. They started walking toward the scene to get a better look. No, wait, they weren't walking. They were running, and they looked angry.

"We'd better get out of here, Trace," Blaine declared, "and quick!"

The twins managed to slide off of the pile of woven baskets and get to their feet.

"I am so sorry, Ma'am," Tracey murmured apologetically to the vendor lady.

"It's okay," the woman responded, managing a smile. "You just surprised me."

The twins were thankful that the vendor had been forgiving, but the guards that were racing towards them looked as though forgiveness was the last thing on their minds. Without pause, the Sassafras twins burst into a full sprint, away from the oncoming guards.

The market that they were in stretched practically as far as the eye could see. There was stall after stall with various goods for sale—everything from produce to live animals, furniture, books, artwork, and more hand-woven baskets filled the narrow streets. People were everywhere, some looking to buy and some looking to sell. Vendors shouted out their prices; buyers answered back with bartering shouts of their own. Old men sat on stools smoking pipes; old women looked carefully over piles of fruit; children played with wooden toys on the ground. The Sassafrases raced for all they were worth through the whole of it all.

They zipped past stalls and shops, dodging carefully but quickly around people and merchandise. Tracey managed to glance behind and saw that the angry guards were hot on their trail.

"These guys are fast, Blaine!" she shouted to her brother, who was just in front of her. "And they are still mad. We have got to find a way to lose them!"

Blaine responded by taking a sharp right and cutting over to a different row of the market. Tracey followed. The row seemed to be narrower than the first, making it hard to run very fast. The guards had seen them cut over, had managed to follow, and had even gained some ground on the twins. Suddenly, the Sassafrases saw something that was of some concern up ahead of them. An old man was pushing a whole cart of chickens straight toward them, and his cart was virtually blocking the entire path. The steaming guards were now only feet behind the twins.

Blaine looked straight ahead. He and his sister were going to have to play chicken with a cartload of chickens. Both Sassafrases pressed forward, straight toward the oncoming cart. The man pushing the load of chickens did not see the two children running his way. He pushed the cart onward, taking up the whole lane. The guards were faster than the twins and were gaining on them. The only two outcomes seemed to be either a head-on collision or being grabbed from behind, but still the twins raced on.

Blaine reached the chicken cart full speed ahead, but instead of crashing into it, he jumped up, put a foot on the cart's front corner, and catapulted himself over the cart. He landed safely in a somersaulting tumble on the ground behind the cart. Tracey followed suit, jumping and landing in similar fashion. The two guards, however, were not so lucky. They both crashed, with a loud smack, right into the front of the cart, creating a small explosion of dust, clucks, and feathers. The twins didn't stand around to see what else happened. They immediately picked themselves up off the ground and raced on through the maze.

After running for another five minutes or so without being pursued, Tracey saw a door that had a 'Sorry We're Closed' sign on it. She pushed on it in desperate curiosity, and it happened to be open.

"Blaine! In here!" she called to her brother.

Blaine looked up and saw that he was following his sister into a shop called 'Raz's Pawn and Antiques.' The two snuck in and closed the door behind them. Both took a deep breath of relief and slid to their seats in the dark room that they now found themselves in.

"That was a close one," Tracey whispered.

"Sure was," Blaine answered. "What a way to start a new subject."

Blaine looked down in his hand and saw that he had managed to hold onto the old tattered backpack that he had found on the awning throughout the entire chase. "Tracey," he chuckled, as he showed his sister, "look what I have."

Tracey just shook her head. "I saw you were carrying that ratty old thing. Why?"

"I don't know. I guess I just never let go of it."

As the twins' eyes adjusted to the darkness of the room, they saw that the shop was fairly small. It had shelves and shelves of old looking things. There were statuettes, bowls, clocks, coins, knives, ornate boxes, instruments, jewelry, and more. The place smelled of dust and was all dark and quiet except for the dim glow of a light coming from a separate room in the back of the shop. The twins stopped their own conversation as they heard the faint buzz of another conversation coming from the direction of that back room.

The children stood up and slowly walked toward the faint sound and light. They had made it to the edge of the open doorway, when they stopped for a closer listen, making sure they stayed hidden. There were two voices. The conversation was high energy but seemed more friendly than heated.

"Of course I know a skull when I see one, Raz," the first voice defended. "Do you want me to break it down for you? A skull is made up of twenty-two separate bones. Eight of them form the cranium, which is the dome-shaped bony box that surrounds and protects the brain from being smashed or damaged. The other fourteen bones make up the facial structure. Only the lower jaw is able to move freely. The bones of the skull meet at jagged edges that line up and lock tightly together, like a puzzle, giving the skull its strength. The skull has openings for the ears, nose, mouth, and eyes. The eye sockets allow the eyes to move freely but also provides them with a protective pocket. There are two rows of teeth, one row anchored in the upper jawbone and one row in the lower jawbone. Humans use these teeth to bite and chew food with. There, pal, I told you I know what a skull is. Now, why would you want to sell me a skull? I came here to see if you have come across the map yet."

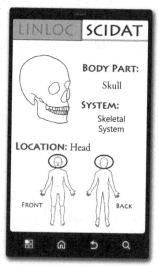

"Snowflake, Snowflake, my friend," the second voice soothed, "you are the smartest man I know. Of course you know what a skull is, but this is no ordinary skull. Take a closer look at the teeth."

There was a long pause and then the man named Snowflake (who the twins were sure had to be their local expert, Larry 'Snowflake' Maru) exclaimed, "It can't be! That's impossible!"

There was another long pause.

"You know what this means, Raz?" Snowflake urged.

"It means that the Legend of the Seven Monks' Tomb is... true."

"The Legend of the Seven Monks' Tomb?!" Blaine blurted out in amazement.

"Shhh!" Tracey voiced.

But it was too late. Blaine had talked too loudly, and now the two men in the back room knew that they were not alone. A tall bearded Ethiopian man walked slowly around the corner of the door frame and spotted the twins. He was wearing dusty boots, cargo shorts, and a long-sleeved denim shirt that had two pockets on the front. He had a satchel that looked to be packed full, and he was wearing a sweaty brown fedora style hat that was hiding his curly salt-and-pepper-colored hair.

"And who might the two of you be?" the man asked.

Blaine seemed to have gone mute after his untimely blurt, so Tracey answered the man's question.

"We are the Sassafrases. I am Tracey and this is my very loud brother, Blaine."

Another shorter man now appeared. He was dressed in overalls.

"How did the two of you get in here? Didn't you know that we are closed right now?"

"Oh, it's all right, Raz," responded the tall man, kindly. "They're just kids. Maybe they're interested in…" the man's sentence stopped abruptly as he spotted the old backpack in Blaine's hand.

"What is it you have in your hand, son?"

Blaine, evidently now fully recovered from his speechlessness, answered, "Ummm, this thing? It's just an old backpack that I found that somebody threw out as trash."

The tall man smiled and laughed. He reached over and gave Raz a friendly slug in the arm. Raz reached up and rubbed his arm; the look on his face showed that he had no idea why his friend was laughing. The taller man looked back at Blaine.

"Would you be so kind as to let me look at that backpack, son?"

"Sure," Blaine affirmed, happily, "but first tell me this, do you happen to be Larry 'Snowflake' Maru?"

The man raised his eyebrows in surprise. "Why yes, son. That's me."

"Dr. Larry 'Snowflake' Maru," Raz corrected. "He's an archaeologist."

"You can just call me Snowflake."

Blaine handed Snowflake the ragged backpack. Snowflake took the bag with both hands and looked at it carefully, like it was a priceless treasure.

"I cannot believe it," he breathed quietly.

The twins had no idea why an old backpack called for such reverence. Snowflake studied the backpack a little longer and then announced, "Let's look at this under better light."

The twins followed Raz and Snowflake as they walked back into the back room. Snowflake carefully placed the backpack under a lamp on a table next to a skull that was already sitting there. He and Raz took seats at the two chairs that were already at the table.

The Sassafrases stood at the edge of the table, both snapping quick pictures of the skull.

"Tell me where you found this again, son," Snowflake said as he looked directly at Blaine.

"It was in a pile of trash on top of an awning out in the market," Blaine responded.

Snowflake nodded, acknowledging Blaine's answer, and then he carefully pulled back the ripped front of the backpack and took a long close look at the lines and writing that were stitched on the inside.

"Today is a good day," Snowflake finally exclaimed. "It's almost too good to be true. Miraculous, even." He looked across the table at Raz. "Well, Raz, my friend, I'll pay you whatever you want for this skull of yours."

Snowflake then looked at the twins. "Blaine and Tracey Sassafras, in my expert opinion, this bag that you have found is very valuable. I don't know that I can even put a price on it."

"Tell you what, Doc," Blaine piped up, trying to make a deal. "You teach us a little something about the skeletal system and we will give you this bag."

A huge smile formed on Dr. Larry 'Snowflake' Maru's face. He stood up from the table. Blaine thought he was going to offer a handshake to seal the deal they had just made. Instead, the tall man wrapped both him and Tracey up in a huge hug.

"You two Sassafrases are like angels to me," he said, happily. "For today it's as though you have given me a gift from Heaven."

Neither Blaine nor Tracey were sure how an old ragged backpack could be like a gift from Heaven, but they were glad that their newest local expert liked them as much as angels. Snowflake released the twins from the hug but kept a hand on each of their shoulders. The big smile was still on his face, and he now asked with a twinkle in his eye, "Blaine and Tracey how would you two like to hear a story?"

Legends and Backbones

A few hours later, Blaine and Tracey found themselves sitting in big comfortable chairs at a cozy little coffee shop called "Raz and Retta's." The place was beautifully decorated. It had been painted in deep oranges, yellows, and blues. It was accented and trimmed out in dark brown wood and had nicely crafted lamps at every small table, casting a warm glow on the coffee shop. Snowflake's friend, Raz, was the owner of 'Raz's Pawn and Antiques,' but he also owned this coffee shop. Raz, who was sitting here with the twins and Dr. Maru now, said his wife, Retta, was the brains and beauty behind this place. They had a seat reserved here at their quiet corner table for Retta too, but she couldn't sit still for very long because she was constantly up and down, striving to make sure that all the patrons were happy. Retta was a beautiful woman with a kind smile, and her personality oozed hospitality. It was no wonder 'Raz and Retta's' was packed full with happily conversing people drinking fantastic Ethiopian coffee. Even the twins, who were only twelve years old and not exactly 'coffee connoisseurs,' were enjoying what was in their cups. More than that, they were enjoying getting to know their interesting local expert better.

"So where did the name Snowflake come from?" Tracey asked, dying to know.

Both Raz and Maru laughed at the question. "It's that silly movie," Raz said.

Maru nodded and then expounded. "When I was younger I saw a movie about an archaeologist that was named after his dog. I loved the movie and wanted to be an archaeologist, but my problem was I didn't have a dog. I only had a skinny little white cat named 'Snowflake.' When my friends found out how much I liked the movie and that I was aspiring to be an archaeologist, they started calling me 'Snowflake' to poke fun. I didn't mind it so much, so the nickname stuck. Today, I proudly go by the name Larry 'Snowflake' Maru."

"Oh, you men. Quit talking about yourselves," Retta quipped, coming back over and sitting down to join them. "Tell these children the story."

Blaine and Tracey sat up a little straighter at the mention of this. They were both very curious about all they had heard back at the pawn and antique shop.

"Remember, children," Retta warned, "When someone attaches the word 'legend' to something, you must listen with ears that know that all they're hearing is not necessarily truth."

Both Raz and Snowflake chuckled at the plainly spoken words of Raz's wife.

Snowflake cleared his throat and pushed his hat up further on his forehead. "Nevertheless," he paused and leaned forward in his chair before he continued in a deep voice, "let me tell you about the Legend of the Seven Monks' Tomb."

The way he said it sent excited shivers down the twins' spines.

"The story starts with King Solomon, of biblical fame. King Solomon built the greatest temple the world had ever seen in the city of Jerusalem. The centerpiece to this temple was a magnificently handcrafted golden vessel called the Ark of the Covenant. This golden ark housed a heavenly loaf of bread called manna, the staff of Aaron, Israel's first high priest, and the original Ten Commandments that were hand carved by Moses on tablets of stone. The Queen of Sheba, from Ethiopia's most ancient royal line, and King Solomon are said to have had a son together. Their son's name was Prince Menelik—the very same Menelik that later became the first emperor of Ethiopia. When the city of Jerusalem was sacked by Pharaoh Shishak of Egypt and Solomon's temple was raided during the night, Menelik and his brave band of Ethiopian warriors secretly hauled away the ark from his father's ravaged temple and hid it in another safe location in Jerusalem.

"The ark remained in Jerusalem until King Nebuchadnezzar of Babylon burned the city to the ground, but, just as before, a descendant of Menelik stole the ark safely away during the night. Legend has it that this brave group brought the Ark of the Covenant to the Ethiopian city of Axum, where it was hidden deep in the belly of a castle made from solid rock. That castle has since been converted to a church, and the ark is said to be still superbly hidden within its solid walls, being continually watched by a select group of monks whose sole vow is to protect the sacred vessel. Snowflake paused his story and looked directly into the twins' eyes, which were wide with wonder.

"But that's not the whole story," he said slowly.

The archaeologist leaned forward even more. "I believe the whole Axum location and rock-hewn church story is just a made-up story. I believe that Menelik's line let this false story leak out to keep treasure hunters looking in the wrong places."

The Sassafrases could barely handle the suspense.

"So where is it?" Tracey asked eagerly. "Where is the real location of the ark?"

Snowflake smiled a wry smile. "This is where the legend of the Seven Monks' Tomb truly begins," he said.

Retta rolled her eyes, like she had heard this story a thousand times and didn't believe a word of it. She silently laughed and got up to make sure everything was going well in the rest of the coffee shop. Her departure did not dampen either of the twins' enthusiasm to hear the rest of Snowflake's story.

"So tell us," Blaine said desperately, "what's this Legend of the Seven Monks' Tomb?"

Snowflake Maru carefully stroked the whiskers of his beard, and then he continued. "I believe that the Ark of the Covenant was actually brought by those Ethiopian warriors to the capital city of Ethiopia, the city of Addis Ababa, which just so happens to be the city that we are sitting in right now. The Legend of the Seven Monks' Tomb says that the ark was hidden somewhere in the royal catacombs, where only Ethiopia's bravest and most noble are buried. The Holy Trinity Cathedral was built over the top of that site, and now the only entrance to the catacombs is hidden somewhere within the cathedral. Seven stalwart monks were chosen to guard the ark in a secret tomb hidden away in the dark corridors of the royal catacombs. Over the years, if one of the seven monks died, he would be buried in a chamber connected to this secret tomb and would be immediately replaced by another monk that was chosen by an Ethiopian emperor in the line of Menelik. These monks were so fiercely dedicated to the royal and divine task that a custom arose that was unique only to them; they each had a symbol representing the ark engraved into their teeth."

Blaine and Tracey shuddered at the thought of having their teeth engraved.

"During the battle of Adwa," Snowflake continued, "which was the first of the Italo-Ethiopian wars, a Special Forces brigade of Italian soldiers, following the directions of their commander, broke into the Holy Trinity Cathedral and found the entrance to the royal catacombs. They had been instructed not to return to their camp without the ark, for it was to be the Italian army's prize of the ages. After days of searching the vast maze of tombs, crypts, and burial chambers, the Italian soldiers were on the cusp of finding the tomb where the ark was being kept. The seven monks were well aware of the soldiers' presence and knew that the Ark of the Covenant was in grave jeopardy of being stolen. But they had designed the ark's tomb for such a moment as this. The monks pulled a hidden lever that brought down a gargantuan wall of stone, effectively blocking the Italian soldiers away from their prize. But what it also did was lock all seven of the monks inside of what would now become their eternal, impenetrable tomb. The Italian soldiers left the catacombs and the cathedral empty handed.

The Ark of the Covenant was now forever locked inside a hidden tomb, with no way in and no way out. But legend says that every one of the Italian soldiers made maps by stitching directions to the Seven Monks' Tomb hidden somewhere in their army gear. They had hoped they could bring back a larger regiment of the Italian army, find the tomb, break into it, retrieve the ark, and win back their honor, but it was not to be so. The Ethiopians, under the command of Emperor Menelik II, decimated the Italians during the Battle of Adwa and drove them from their land. Those Italian soldiers were never seen or heard from again. Their maps, stitched inside their army gear, were gone forever."

Dr. Larry 'Snowflake' Maru added another pause to the story as he reached inside his leather satchel and pulled out the old backpack that Blaine had found. He carefully placed it on the coffee table.

"Gone forever, that is, until you, Blaine Sassafras, found this in a pile of trash today."

Blaine's mouth dropped open. "You mean," Blaine stammered, "that backpack that I found is..." He was so awestruck he couldn't even finish.

"One of the soldiers' lost bags, with a map stitched on the inside leading to the Seven Monks' Tomb," Snowflake declared, finishing the sentence that Blaine couldn't.

The Sassafras twins simply couldn't believe it. Blaine looked over at Tracey with a look on his face that seemed to say, "See? I told you that bag wasn't trash."

Now Tracey was just as happy as Blaine and Snowflake that she hadn't talked Blaine into tossing the bag. "It really is miraculous," Tracey thought to herself, but she still had questions.

"OK, but what about that skull that you bought from Raz? What is so important about it?" Tracey asked.

Snowflake grinned and reached into his satchel again. "I'm

glad you asked," he said, pulling out the skull and setting it on the table next to the now very significant Italian army bag.

"Do you two remember how I said the seven monks had a symbol engraved on their teeth?"

Blaine and Tracey both nodded, vividly remembering that part of the story.

"Well, then," the doctor of archaeology beckoned, "take a closer look at the teeth in this skull."

The Sassafras twins slowly leaned forward and gazed, through the dim light cast there on the wooden table by the lamp, at the teeth of the skull. Now that they knew what they were looking for, they both saw it immediately. On every tooth was an engraving of what looked like wings over a box.

"It's the symbol of the Ark of the Covenant!" Blaine exclaimed.

"It's hard to see at first," Tracey added excitedly, "because the engravings are the same color as the teeth, but they are there."

The wheels in Blaine's mind were cranking at full speed. "So, by finding this skull, we know that the legend is not only true, but that there is a way into the tomb!"

"Correct, young man," Snowflake confirmed, "because if a skull from one of the monks has been found outside the tomb, it means the Seven Monks' Tomb wasn't impenetrable after all. The only possible explanation is that someone found a way into the tomb and retrieved the skull."

"Or someone found a way out," said Raz, who had been quiet up to this point.

"Out?" asked the twins in unison.

"Yes," Raz nodded. "Maybe the monks weren't locked in after all. Maybe one of them found a way out."

"Regardless," Snowflake dismissed, "today we found a skull

that proves the existence of the Seven Monks' Tomb, and we also found a map leading to that same tomb. So, tomorrow," the tall Ethiopian said, smiling, "we set out together to see if we can find the Ark of the Covenant."

"Wow," Blaine and Tracey thought. "What an adventure!"

"Aren't you forgetting something?" Retta asked, as she returned to the table. "Neither one of the two of you..." she pointed to Snowflake and her husband with a big smile on her face, "has a backbone"

"No backbone?" Raz said, somewhat defensively. "What do you mean?"

"Both of you are scared of anything that crawls, hisses, or slithers. Those catacombs you are planning on going to tomorrow are bound to be overflowing with all kinds of creepy critters."

Both men gulped; maybe Retta was right.

"We have backbones," Snowflake said, trying to defend himself and his friend.

He pulled out a sketch pad and a charcoal pencil and started drawing. "The backbone, otherwise known as the 'spine,' provides a protective tunnel for the spinal cord. It stretches from the base of the skull to the bottom of the pelvis. It is a strong and flexible rod that helps keep the body upright."

"You know what I meant," Retta chuckled at her friend. "By no backbone, I meant you two are scaredy cats."

Snowflake Maru ignored Retta and kept talking and sketching. "The spine consists of thirty-three vertebrae, with the lower five being fused together. Each vertebra has a central portion called the centrum, which helps to bear the body's weight; the upper vertebrae also have a hole in the center of their centrum, which allows the spinal cord to pass through. Each vertebra also has a disk of cartilage in between, which cushions the space between each centrum. This allows for tiny movements in between each vertebra.

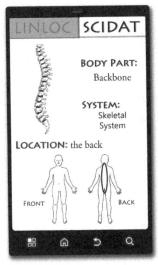

The movements combine together to allow the body to bend and twist in many directions. Each vertebra also has a spinous process, which is the bumpy part of the backbone. It gives space for the muscles to attach."

The twins were so glad their local expert was so knowledgeable. This information about the backbone was just what they needed for their SCIDAT data.

"The backbone," Snowflake continued, "is divided into three main sections. The top section, or cervical spine, contains the vertebral bones of the neck. The middle section, or thoracic section, contains vertebral bones of the upper and middle back. And the lower section, or the lumbar section, contains the vertebral bones of the lower back, including the five fused bones of the sacrum."

Dr. Maru finished his sketch and held it up for everyone to see, especially Retta. The twins used their phones to snap a picture of Snowflake's drawing.

"Very nice," Retta said, standing up again to go check on the happenings of the coffee shop.

"It's getting late, you two," she announced to the two men. "If you all are really going tromping off tomorrow to find this treasure of yours, you'd better get some rest."

"It's not only about the adventure and the treasure," Snowflake smiled. "It's also about the science. My greatest desire is not for riches in my hands, but riches in my mind."

"That's one of the many reasons I'm glad you are friends with my husband," Retta laughed. "Now, you two get to bed and make sure you show these children where they can sleep."

Snowflake snatched up the skull, the sketch, and the army backpack back, put them into his satchel, and then stood to leave.

"There are some spare rooms on the second floor here at the coffee shop," Raz shared as he and the twins stood up. "We'll sleep here tonight, and then we'll head to the cathedral first thing in the morning."

"Also," Snowflake mentioned to Blaine and Tracey, "we will talk more about the skeletal system tomorrow. That is a part of our deal, right?"

The Sassafrases nodded.

"Like, for instance," Maru offered, "did you know how the skeleton is a bony framework of 206 bones that support the body, and that the axial skeleton or the central core of the skeleton is the skull, backbone, and ribs? The skeleton, as a whole, is strong, yet flexible, and protects important organs and anchors muscles. Without the skeleton, our bodies would just be a floppy messes. The skeletal system doesn't just include bones, but also the tissue that connects them, such as tendons, ligaments, and cartilage. If we actually find the Seven Monks' Tomb and get inside, we will be able to see and study some skeletons firsthand."

"That sounds great," Blaine proclaimed. "This is going to be awesome!"

As the four headed upstairs to the spare rooms, Tracey had one more question for the archaeologist. "So, are you really scared of anything that crawls, hisses, or slithers?"

A little bit of blood drained from Snowflake Maru's face as he answered. "Oh, yes, I am. Very scared."

THE SASSAFRAS SCIENCE ADVENTURES

Those twins were fairly smart and definitely resilient, but they weren't very observant. It had always been so easy for him to sneak up on them and then hide in plain sight. Here he was, sitting one table over from where they had been. He was wearing plain clothes so he wouldn't stand out, and he had donned a brown fedora, like many men in this place were wearing. His fedora had served its purpose perfectly: to hide his brow and conceal the fact that he had no eyebrows.

He had just heard every word the long-winded archaeologist had said. He hated the fact that the children had just gotten information on backbones, but he would just have to let that go, so that the plan he had for Ethiopia would work. Good thing they had just given away their intended location for the night. This was going to be much easier than he had first thought. He gave a nod to the three men he'd been sitting with, and they all four got up. He didn't much like these guys—they were big and mean and rude, but they would serve his purpose perfectly.

They were men he'd met earlier in the day, at the black market. Their expertise was in trading, buying, and selling stolen goods. He could almost see them salivating as they had listened to that Snowflake fellow talk about the legend of the Seven Monks' Tomb. These men now wanted that ark badly, and he was pretty sure they would do anything to get it, but his prize was not the ark. He could care less about some dusty box. What he wanted was much more important to him. He wanted to stop those children and get his revenge on Cecil Sassafras.

He tossed some money down on the table to cover their coffee bill; then he and his three shady acquaintances walked out into the dark night. They would wait for just a few more hours to make sure the two men and the twins were asleep. Then they would break into their room and steal…

CHAPTER 3: THE CATACOMBS

Blocks and Bones

The coffee shop had been closed for over two hours now. All the patrons had left. All the lights and machines had been turned off, and now there was not a sound or movement coming from Raz and Retta's. The streets were dark and silent as well. It was time. He gave a signal with his hand and stepped out from his hiding spot, his three scoundrel companions following right behind him. They hadn't spent a lot of time planning this out, but it should be enough.

They easily made it across the street without being detected and swiftly scampered around the corner of the building to the backside of the coffee shop. Upon reaching the back entrance, one of the black market fellows pulled out a pair of bolt cutters and smoothly snipped the padlock that was securing the back door. The four quickly filed into the building like silent shadows, and they found themselves in the coffee shop's kitchen. They stood in silent darkness for a moment, making sure that all remained quiet and still.

When he was satisfied their intrusion had gone undetected, he gave the signal. The four of them carefully walked out of the kitchen into the sitting area. Now they just needed to climb the stairs and find the spare rooms. He grinned there in the darkness. So far, everything was going just as planned.

On the first night of their summer, while studying zoology, the Sassafrases had slept outside on the ground behind a thorny acacia barricade, hoping not to get eaten by predators. This night, the first night of their anatomy leg, was faring much better. The twins now found themselves sleeping in a quiet room, in comfortable beds, without a care in the world. Dr. Larry 'Snowflake' Maru was in a separate room close by, as were Raz and Retta. The only possible sound the twins could hear would be the light snoring coming from Dr. Maru's room, but they couldn't hear it because they were sleeping too deeply.

He went up the stairs first, followed by the three black market dealers. They took the steps one at a time, testing each for creaks before they put their full weight on it. After they reached the top of the staircase with virtually no sound, the four stealthily continued down the hallway. There were only four doors up here. He suspected that these were most definitely the bedrooms that he had heard Raz speak of earlier in the evening. Each man took a door and started to pick the locks as quietly and meticulously as possible.

Blaine and Tracey both lay perfectly still in their beds, sleeping on comfortable mattresses and covered up by cozy blankets. Each twin was enjoying such deep sleep that neither was even dreaming. They were just breathing slowly and softly.

With a click, the lock that he was working on disengaged, and the door opened slightly. He held the door there, just barely open, and waited to see if the slight sound he'd caused had disturbed anyone in the room. Complete silence. Good. Now he stood patiently and waited for his three companions to get their doors open. The plan was to enter the different rooms all at the same time. He heard a click, then another. He waited, almost holding his breath, to hear that last lock being undone. Click. There it was. Now to silently enter the rooms.

"What was that sound?" Tracey thought to herself, still very much asleep. "Was that Blaine? Why is he making so much noise? Doesn't he know we need our sleep?" Whatever had made the noise fell silent and Tracey drifted back off to sleep.

Quiet footsteps across the floor; he had practiced this kind of sneaky walking so much recently that he was now able to manage virtual silence as he walked. He hoped his black market friends were being as quiet as he was. The room that he had broken into had two beds. "Probably the twins' room," he thought. More than likely, the skull and the backpack they were looking for were in the room of that archaeologist. But he would look around in here anyway. When they found the two items that they were seeking, the plan was to leave as quickly and quietly as they had come in. He'd told the three dealers not to harm anyone in this thieving process, but he was fairly certain that they would not stick to this part of the bargain.

Bang!

Oh, no! What was that? He had just knocked something over, causing an excruciatingly loud noise.

Bang!

"What was that?" Tracey thought, immediately awake. She looked around the room for the noise's source. There it was. He had done it. Blaine had knocked over the lamp that stood on the table between their beds.

"Blaine," Tracey murmured. "What are you doing? I'm trying to sleep here."

"Sorry," Blaine muttered, apologetically. "That was an accident. But how can you still be sleeping? Aren't you excited about trying to find the Seven Monks' Tomb today?"

"Of course I am," Tracey mumbled, still groggy. "But what time is it?"

"I don't know," Blaine responded. "But the sun is already starting to come up. C'mon, Tracey, let's get this show on the road!"

He stood completely still and silent in the pitch-black room. He still wasn't sure what he had knocked over, but whatever it was had made a loud enough noise to wake more than just the twins. He still didn't hear any movement coming from the two beds. That was strange. Suddenly, the light to the room flashed on. He stood there stunned in the illumination. The room was empty. No twins, no archaeologist, no skull or backpack, only a group of armed Ethiopian policemen standing in the doorway.

THE SASSAFRAS SCIENCE ADVENTURES

The twins gawked in awe as they now walked with Snowflake Maru and Raz through the entrance of the Holy Trinity Cathedral. It was a breathtaking place, with tall pillars, magnificent archways, and high ceilings. This indeed looked like a fitting location for something as sacred as the Ark of the Covenant to be hidden underneath. Raz's cell phone rang.

"Hello," both Sassafrases heard him answer, as they wandered off, continuing to look around at the cathedral's wonders. Here was the tomb of Amha Selassie, Ethiopia's last emperor. Here was something called the Holy of Holies and here was another ark—not the ark they were looking for. This ark sat right out in the open, displayed for all to see. It was called the Ark of Saint Michael the Archangel.

"You are absolutely not going to believe this!" Raz exclaimed, grabbing back the twins' attention. "Retta was right! Those shady fellows that she spotted last night sitting a table over from us were indeed dealers from the black market."

"And that's not all!" he said, in an excited voice that was

much too loud for this quiet place. "They did break into the coffee shop!"

Blaine and Tracey couldn't believe it. Last night, after they had walked up to the upstairs rooms of the coffee shop, Retta had rushed up after them, a little worried. She'd suspected that a small group of men sitting in the coffee shop were dealers from the black market. Not only that, she was sure they had been eavesdropping on the four's conversation as Snowflake had shared every detail about the skull, the Italian army bag, and the Seven Monks' Tomb. Retta was worried that these men might try to break in and steal the artifacts, so she had called the police and secretly moved her husband, Dr. Maru, and the twins to a small guest house a few blocks away.

"Wow! What amazing intuition my wife has," Raz remarked.

"So the police captured them and took them into custody?" Tracey asked.

"Yes, the police did manage that," Raz answered. "Retta also told me that something very strange happened at the police station. Three of the four men captured were known criminals and are still in custody, but the fourth was unknown. They said he didn't have any eyebrows, but before the police could properly identify him, he just disappeared."

"Disappeared?" the Sassafrases asked in unison.

"Yes," Raz confirmed. "Disappeared into thin air."

Blaine and Tracey were speechless. The Man with No Eyebrows was back. They hadn't seen him since Peru, but now, he had been spotted again in Ethiopia, and he was still trying to ransack their scientific adventures. They were sure that he had his own invisible zip lines and that he was using those lines to stalk them, but why? Why was he trying to stop them and ruin them? They just had to get this mystery figured out.

"OK, everyone," said Dr. Maru, who had pulled his own

disappearing act. He now rejoined the group, accompanied by a gentle looking priest, fully decked out in his priestly robes. "This is Brother Eskinder, and he is a custodial priest here at Holy Trinity Cathedral."

The priest nodded and waved his hand at this introduction.

"He was giving me some bad news and some good news. The bad news is that no one is allowed in the royal catacombs."

The twins and Raz all exhaled sadly, deflated a bit.

"But," Snowflake continued, "the good news is that after I showed Brother Eskinder the skull and the backpack, he said these discoveries called for an exception, and that he would show us the supposed entrance to the underground tombs."

"Yes," Brother Eskinder said happily. "I have heard the Legend of the Seven Monks' Tomb, and I have always wondered if it was true. The four of you should definitely have the opportunity to search for the tomb. How amazing would it be if this cathedral that I love so much is housing something as sacred as the Ark of the Covenant? It would be truly miraculous."

The priest paused and looked over the group of four adventurers. He then turned and waved for them to follow.

"This way," he motioned.

A few minutes later, the four found themselves standing at the top of a dark stone staircase. Brother Eskinder had shown them a small door in the floor in a corner behind the Emperor's tomb. It had been covered by a heavy wooden table. The group had moved the table, opened the door, and now here they stood.

"The best of luck to you," said the Ethiopian priest. "If you find anything down there, please come back up and inform me."

Larry 'Snowflake' Maru tipped his hat to the priest, clicked on his flashlight, and then started slowly down the steep staircase. He was closely followed by the twins, with Raz bringing up the rear.

They each had flashlights of their own. When the group reached the bottom of the staircase, they looked back up at Brother Eskinder and waved. The priest waved, and then slowly closed the door, leaving the group alone down in the legendary royal catacombs. They had the combined light of their flashlights, but it was still very dark.

"Well, my friend, you actually made it!" Raz affirmed to Snowflake. "You are standing in the royal catacombs. The Ark of the Covenant is now closer than ever."

Snowflake smiled and nodded, but then his face grew serious. "Yes, we are closer than ever, but still we are so far away. These catacombs are known to be a confusing maze of tunnels and passageways, tombs, and crypts. Even with a map, it will be difficult to find the Seven Monks' Tomb."

The doctor reached into his satchel and pulled out the old tattered Italian army backpack that Blaine had found the day before in a pile of trash. It was hard to believe that this worn piece of gear held the map to the prize they were seeking: the Ark of the Covenant. The Ark was the golden vessel that had once been housed in the original Holy of Holies in Solomon's temple in Jerusalem. Snowflake carefully turned the backpack inside out, plainly revealing the map that had been stitched there long ago by an Italian soldier.

All of them shone their lights on the bag as Dr. Maru studied the map. Neither Blaine nor Tracey could make any sense of this map. It looked like just a bunch of random lines of thread, with a border of some words in a language they couldn't read, but this was not the case for Snowflake Maru. He could read Italian and seemed to understand the lines.

"If I'm interpreting this map correctly," Snowflake stated, "there are two main levels to these catacombs. We are here," he said, as he pointed his finger to a spot on the map, "on the top level, but we need to go down to the lower level and find this spot here."

Snowflake removed his flashlight beam from the map and started using it to look around the dark room they were in now. It was a room of decent size. The walls and ceiling were all covered with beautiful stone carvings depicting different saints, Ethiopian nobility, and battle victories. It was amazing that such masterful handiwork would be hidden here underground where most people would never see it. The flashlight revealed that there was only one passageway leading from the room.

"Well, it looks like there is only one way to exit this place," Raz announced, walking toward the passageway's opening.

The Sassafrases and Larry followed the antique dealer. The passageway was long, dark, and narrow. The carvings they'd seen in the entry room continued into the long hallway. They walked nearly fifty yards, and then their flashlights revealed something they hadn't expected—a dead end.

"That doesn't make any sense," Snowflake remarked. "If this was the only exit, why would it be a dead end?"

The doctor turned around and shone his beam back in the direction they had come from. What his light revealed sent the group into collective shock. A stone wall was rising up from the ground and slowly blocking the entrance.

"Run!" Blaine shouted and jumped out to lead the way.

The twelve-year-old boy sprinted toward the narrowing gap, his three companions running close behind. Blaine got to the shrinking entrance in no time, but the rising wall was already over his head. Blaine jumped up and grabbed the top of the upward moving wall. If he could quickly hoist himself up, there could still be enough time to wiggle over and through without getting smashed. Blaine did a sort of pull-up and kicked his feet off the wall to give himself a boost. He managed to get his torso on top of the rising hunk of stone. Could he make it through?

"Blaine!" the boy heard his sister scream. "You are not going to make it!"

A wave of understanding swept through Blaine's mind, and

he realized his sister was right. He quickly slid his upper body back off the top of the ever-rising wall, his back scraping against the stone ceiling as he did. He was now hanging by his fingertips.

Blaine let go right before the wall slammed shut and he landed with a thud on the hard ground. Everyone was relieved, yet discouraged. Before any of the four could truly comprehend what had happened, the floor began dropping out from under them. Not all at once, but one step at a time, literally. The stone floor was dropping quickly down, forming a stone staircase. But as this new development came as a sudden surprise to them all, none in the group were able to actually use the steps properly. Instead, they all fell as a jumbled blob in a tumbling heap down the staircase. Their flashlights flew out of their hands and clanked down the stairs around and behind them, sending abrupt flashes of light ricocheting around the corridor.

Slam! Bang! Crash! Thud! All four finally came to a painful stop at the bottom. Then, just as soon as they had dropped off, the stairs rose back up, effectively trapping the four in whatever hole they had just fallen into.

Tracey managed to pick herself off the ground first. She reached for the flashlight that had landed closest to her, but she froze as she reached because she saw what was now being illuminated in the beam of the fallen flashlight—a skeleton.

Tracey didn't mean to, but she let out a little scream. Dr. Snowflake, who was now picking himself up off the ground, also saw the skeleton.

"No need to be scared, my dear," he insisted, trying to encourage the girl. "There is nothing a skeleton can do to hurt you."

"I know," Tracey said, weakly. "It just frightened me a bit, that's all."

Snowflake picked up the flashlight and used it to get a better look at the skeleton.

"Looks like every bone is here," he said. "The skull is complete. The long bones of the arms join the central core of the skeleton at the shoulder; the long bones of the legs join the central core of the skeleton at the hips. The bones of the arms and legs along with the shoulders and hips make up the appendicular skeleton."

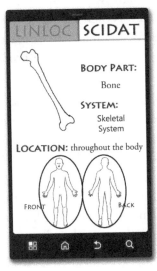

Dr. Maru stepped in for an even closer look. "There are four main types of bones. The flat bones, which give protection and provide surfaces for muscle attachment, such as ribs or shoulder blades; the short bones, which are knobby and nugget-shaped, such as ankles or wrists; the long bones, which are longer than they are wide, such as arm and leg bones; and the irregular bones, which have complicated shapes, such as vertebrae."

Snowflake pointed to different parts of the skeleton as he spoke. "The smallest bones are the three ossicles in your ear, known as the hammer, anvil, and stirrup. The largest bone of the body is the thighbone, or femur. It supports the weight of the body when you stand, run, or jump."

"What about when you fall?" Blaine groaned, finally getting up after the group's tumble. "I think I may have broken my femur."

Snowflake shined his flashlight in Blaine's direction. "Nope. I think you are fine, son. You are standing up, so you didn't break your femur. If a bone is broken, it can be set back into place and over time, new bone will grow to rejoin the fractured pieces. The new joint will be stronger and visible on the bone."

Blaine picked up a flashlight and then hobbled over and joined his sister and Snowflake. He took his phone out of his backpack and snapped a picture of the skeleton that the two had been looking over. Tracey took a picture too.

"Bones are living organs with their own cells and blood supply," Larry continued with the facts. "They have an outer layer of compact bone cells that surround a layer or lighter, sponge-like bone, which is filled with jelly-like bone marrow. Compact bone is made up of bone tubes that are bundled together, making it very strong. Spongy bone has a honeycomb structure made up of spaces and bony strands. All of this works together to make living bones five times stronger than steel."

"The spaces in the spongy bones are filled with bone marrow. There are two types of bone marrow found in the center of the bones: the red marrow and the yellow marrow. The red marrow is responsible for making red blood cells, and the yellow marrow is responsible for storing fat. The largest area of the bone, which is usually found at the joints, consists mainly of spongy bone. This allows the bone to bear weight without bending. The outer layer of hard bone prevents the spongy layer from being squashed."

"That's all well and good, my friend," grumbled Raz, the last one of the group to get up off the stone floor. "But where are we? What happened? Why did Blaine here almost get squashed, and what is this place we have fallen into?"

"It looks like one of the many burial chambers that we will more than likely come across down here in the catacombs," Snowflake answered. "But I don't know why we fell into this place like we did. It is almost like our dear friend, Brother Eskinder, set us up for this."

"And now we are trapped," Raz added.

The group looked around the burial chamber. It definitely looked like Raz was right. There were no doorways here or passageways leading from this place. It looked like a room made for a skeleton, with no obvious way out.

Suddenly, the group heard a loud hissing noise coming from the direction of the skeleton. All four adventurers froze completely still except for their racing heartbeats and their shaking hands. The flashlights they were holding were all pointed at the skeleton laying there on its final resting place—a shelf built into the wall a couple of feet off the ground. The light danced around a bit because of their shaking hands. However, they could see nothing out of the ordinary happening with the skeleton.

Jumping Joints

Hiss! There was the sound again! So loud and clear that it caused the flashlight to drop right out of Raz's hand. Snowflake took a step back; Blaine stood still, but Tracey took a step closer to the hissing skeleton.

"Tracey!" Blaine exclaimed. "What are you doing? Are you crazy? Get away from that thing!"

Tracey ignored her brother. She knelt down and used her light to take a closer look. Hiss! Everyone jumped at the sound except for the Sassafras girl.

"It's not the skeleton," Tracey quipped.

"What?" Blaine asked.

"It's not the skeleton that's making the hissing noise…it's the uromastyx."

"The uromastyx?" Blaine said, puzzled.

But then the confusion on his face turned to a smile as he recalled his zoology. "Uromastyx! Also known as a spiny-tailed lizard. Oh, man, I love these little guys!"

Blaine knelt down next to Tracey and added the light of his flashlight to hers. Tracey was right. There, behind the skeleton, stood a spiny-tailed lizard. The reptile was hissing and swinging its spiked tail, and he wasn't alone.

"Guess what?" Tracey spoke excitedly. "It looks like there is an exit tunnel down here behind the skeleton. It's packed full of lizards, but we found a way out of here! Isn't that great?"

The twins turned back toward the two men, expecting them to be excited about this good news, but both men had frozen looks of terror on their faces. Blaine looked at Tracey.

"Oh, yeah. Don't you remember what Retta said? These guys have no backbone. They are scared of anything that crawls, hisses, or slithers."

"But Dr. Maru wasn't even a little bit scared of the skeleton. How could he be afraid of a little lizard?"

Blaine just shrugged at Tracey's question.

The twins looked back at Snowflake and Raz, then back at each other. Each knew what the other was thinking. They were going to have to lead these grown-ups through the dark, lizard-infested tunnel.

It was no easy task, but they were accomplishing it. The Sassafras twins were nudging and encouraging Larry 'Snowflake' Maru and Raz through the small, reptile-filled passageway. It had taken both men a long time to build up enough courage to enter the tunnel, but here they were now, crawling on their hands and knees through the dark, around and over hissing spiny-tailed lizards. Tracey was in the front, leading the way. Blaine was in the back, making sure neither of the men chickened out and tried to turn back.

"OK. It looks like we only have about ten feet left to go," Tracey was saying in an upbeat voice. "You guys can do it!"

Both men had their eyes shut and were humming to themselves to try and drown out the lizards' hissing. Blaine was pretty sure he could hear one of them crying. Oh well, they were almost to the end of the tunnel. Tracey jumped out first, quickly followed by Raz and then Snowflake. Finally, Blaine crawled out of the exit tunnel and stood to his feet, joining the other three. Raz opened his eyes with a relieved look on his face, but Snowflake's eyes remained shut, and he was still humming. Raz backhanded his friend in the chest, embarrassed.

"We are out, man," he declared. "We made it. You can open your eyes and stop humming."

Snowflake grew silent and slowly opened his eyes, as if he was scared of what he might see, but there were no more lizards. The Sassafras twins had led them safely out of the cramped tunnel. Raz hung his head a little and pointed at his friend's face.

"Snowflake...you have some tears, umm I mean sweat, under your eyes and running down your cheeks."

Snowflake quickly reached up and wiped the 'sweat' off his face.

"OK. Where are we now?" the doctor remarked, trying to recover and regain some respect from his friends. "It looks as though we are in a long hallway of some kind."

He pulled out the stitched map and studied it once again. After only a few seconds, he felt like he knew right where they were. "I believe we are on the lower level of the catacombs now, and if we head this way," he pointed to the left, "we should be going in the direction that we need to go."

Snowflake Maru kept the map out to keep their bearings correct as he began leading the way down the dark hallway. The twins waved their flashlights around as they walked, amazed at all the intricate stone carvings that they were still seeing down here. The hallway led to a "T" and Snowflake looked at the map again, deciding to take a right there. The twins and Raz followed.

Minutes turned into hours, and the hours started stacking up, but still the three followed the archaeologist. They trusted his judgment, but they knew that even with a map, making their way through this maze of passageways and tombs was going to be difficult. They took rights and lefts. They walked up and down stairs and through endless narrow passageways. They backtracked a few times and walked in some circles, but all were certain that they were making progress.

Occasionally, Snowflake would read something that was written on the catacomb's walls: names of nobility buried within the tombs and crypts, stories of courageous Ethiopian heroes, and many accounts of victories in battle. The stone carvings had continued throughout their underground traversing, giving images to some of the things written.

Finally, after what felt like an entire day of walking, the group arrived at what Snowflake thought was the Seven Monks' Tomb.

"This is it?" Raz was saying, somewhat skeptically. "This is where the Ark of the Covenant is hidden? It is nothing but a plain stone wall at a dead end. There isn't any markings on it or anything."

"Exactly," Snowflake asserted confidently. "A wall like this would confirm the legend. When the seven monks sensed that the Italian soldiers were close to finding the ark, they pulled a lever that sent a solid wall of stone crashing into place, blocking them in and the soldiers out. This has got to be that wall! It has to be! This is where the map led us, is it not?"

"Well, if it is the wall that is blocking the Seven Monks' Tomb, how do we get it open?" Blaine asked. "If the Italian soldiers couldn't figure it out, how are we going to?"

"We have something the soldiers don't have," answered Snowflake. "We have the skull, which means we have the knowledge that there is a way in and out of the tomb. If the legend is true, and the monks did use a lever to release the wall downward, then more than likely there is also another lever hidden somewhere called a fulcrum release lever that will pull the wall back up."

The twins didn't really have any idea what Snowflake was

talking about, but they could see that he was getting very excited, so that made them excited. What if they really were separated from the Ark of the Covenant by just one wall? The thought was mind-blowing.

"So now," Snowflake jumped into action, "we look for that fulcrum release lever."

"What will it look like?" asked Tracey.

"I don't know," was Snowflake's unexpected answer. "It may be big, it may be small, but it should stand out somehow."

The archaeologist began scanning his eyes across the unmarked wall. Blaine and Tracey started looking around on each of the side walls, and Raz studied the floor.

"It may be big, it may be small, but it should stand out." Tracey thought over the description that Snowflake had just given. It wasn't much to go by, but she was determined, just like everyone else, to scan every inch of these surfaces until they found this fulcrum thing. After looking awhile, they all concluded that this release lever probably wasn't big. If it was, one of them would have seen it by now.

Tracey's eyes were starting to get tired. She was sure she had looked over her wall in its entirety, but she would look again, re-checking what she had already looked at three or four times.

Wait. What was that? There, in the bottom corner of her wall. Were her eyes deceiving her? Or was that the Ark of the Covenant symbol? The same symbol that was engraved on the teeth of the skull. The box shape with the wings over it. Except this symbol was a little different. It had three dots inside the box. The symbol was down close to the floor and partially covered in dust, so Tracey knelt down for a closer look. She wiped the dust away with her hand. Her eyes had not deceived her. It was the symbol.

"Dr. Maru," Tracey called. "Take a closer look at this."

Snowflake bent down and looked at the symbol Tracey had

found. He gasped, then smiled, then frowned, and began stroking his beard like he was in deep thought. He paced back and forth across the corridor, without saying a word. He would stop, kneel down, and look at the symbol. Then he would stand and start pacing again. He repeated these steps, seemingly endlessly. Then, all at once, he stopped. His mouth dropped open.

"No! That couldn't be it, could it?" Snowflake bent down and looked at the symbol again.

"Upper molars," he said, plainly.

"Upper molars?" Raz said, bumpuzzled. "What do upper molars have to do with anything?"

Instead of replying, Snowflake got up and walked over to his satchel, which he had set down. He pulled out the skull. He studied it carefully, and then he began pulling at one of the teeth. The Sassafras twins looked at each other with confused faces. What was the archaeologist doing? The tooth that Dr. Maru was pulling popped out, and he held it up for his three friends to see.

"The upper molars have three roots. And if I'm not mistaken..." Snowflake stated, as he took the tooth over and knelt down next to the ark symbol. "...the roots on this upper molar from the skull that came from inside the Seven Monks' Tomb should fit perfectly into this symbol that Tracey found."

Snowflake lined up the three roots of the tooth with the three dots on the symbol. Now the twins saw that the three dots were actually three holes. The tooth fit perfectly. Snowflake pushed in and then twisted. As soon as he twisted the tooth, the small ark symbol turned and a loud cranking noise sounded from somewhere within the wall. Then, to everyone's amazement, the unmarked wall started to slowly rise.

"The Seven Monks' Tomb," Raz gasped. "This is it, and it's opening."

The twins stood there in speechless awe. Were they about

to find the long lost Ark of the Covenant? The huge heavy solid stone wall continued to inch upward. It creaked and cranked until it came to a complete stop, concealing itself up in the ceiling. What the four now saw clearly in front of them was astonishing. None spoke a word. They just stood there with wide eyes and mouths open in wonder. After a long time, Blaine was the first to speak.

"They are all here. All seven of them."

Snowflake nodded. "But it's not. It is missing."

The room/tomb that was in front of them had seven seats inside it; each of those seats had a skeleton sitting on it. They were the skeletons of the seven monks. The skeletons and the seats were all facing in toward a short platform, but there was nothing on the platform. It was empty. It is where the Ark of the Covenant should have been, but it was gone.

Snowflake slowly stepped into the tomb and started carefully examining everything: the platform of the missing ark, the stone seats, and the skeletons with engraved teeth. Suddenly something struck Blaine as strange.

"Wait a second," the twelve-year-old said. "All seven of these skeletons have skulls. None is missing. So how did we get a Seven Monks' Tomb skull?"

Snowflake looked at Blaine. "We all have questions, son," he answered kindly. "But it doesn't look like we are going to get very many answers today."

The archaeologist paused and looked around the tomb. "But one thing I can do for you is finish teaching you about the skeletal system. You gave me the backpack with the map, so that was our deal, right?"

Blaine nodded.

"Seeing how these skeletons are in the sitting position reminded me how I haven't yet told you and Tracey about joints yet," Snowflake said. "Joints are movable points that allow the skeleton to move. The body has over four hundred joints. There are different types of joints that allow for different kinds of movement. Hinge joints work like a door hinge. They allow for bones to move up and down but not side to side. Examples of these are the knee and elbow joints."

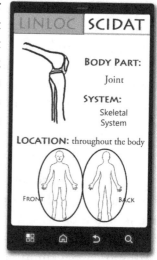

Snowflake pointed to the skeletons as he spoke. "Ball and socket joints allow for movement in many directions. A round end of a bone fits into a cup-shaped socket of another bone. Examples of these are the shoulder and hip joints. Condyloid joints, which are found in the fingers and toes, are oval ball and socket joints that allow for the fingers to swivel but not rotate. Pivot joints allow for side-to-side movement where one bone pivots around another one. Examples of these are the atlas and axis bones in the neck of this skeleton."

"And, finally, we have gliding joints. These allow small sliding movement between two bones. Examples of this are the wrist and the knee caps."

Both Sassafrases took pictures of the different joints as the archaeologist spoke.

"And just for your knowledge," Snowflake continued. "All that remains of a skeleton are the bones. So there is no cartilage or ligaments left, but let me tell you about these anyway, because they are a very important part of the skeletal system. Cartilage is a tough rubbery material that supports different parts of your body, like your ears and the tip of your nose. Ligaments are strong elastic tissues that secure and hold bones and muscles in the body. The ends of each bone at the joints are covered with cartilage, which allows them to slide over each other as they move, and the ligaments hold the two bones in place within the joint."

Dr. Larry 'Snowflake' Maru stopped there and sighed. "I think I have now covered the major parts of the skeletal system. I just wish you had a priceless artifact to go along with your new knowledge."

"No worries, Snowflake," responded Tracey, happily. "You have given us more than enough. Thank you for letting us share this adventure with you."

"Yeah, pal, no sweat," Raz added. "The Legend lives on, and I know you'll keep looking and searching.

Snowflake smiled and nodded. "Yes, I will."

It wasn't until the early hours of the next morning that the group re-emerged from the depths of the royal catacombs. They had retraced their steps and then found another way out. Snowflake and Raz stepped out first into the nighttime air, followed by the twins. As Blaine reached back to close the heavy door they'd come out of, he happened to look up. There, looking down at them from a dark second floor window, was Brother Eskinder. The sight was startling. Blaine got the other's attention and pointed up toward the window. Brother Eskinder was holding a candle, which was dimly illuminating his face. He was smiling. They hadn't seen him smile before. And now, on every single one of his teeth, the group could clearly see the engraved symbol of the Seven Monks' Tomb. Suddenly the candle went out, and Brother Eskinder disappeared into the darkness.

CHAPTER 4: OFF TO SYDNEY

Breathe, Just Breathe

Both Blaine and Tracey finished the last sip of delicious coffee and set their cups down on the table. Blaine sighed, leaned back, and wiped his mouth. Tracey snuggled down into her chair as well. Besides being super cozy and comfortable, Raz and Retta's coffee shop had some really good joe. Last night, their sleep had been short, but it had been deep. With no more threat of black market dealers, they had actually gotten to sleep in the upstairs rooms of the coffee shop.

Dr. Larry 'Snowflake' Maru, Raz, and Retta had all joined the twins for morning coffee. Retta was back in the kitchen somewhere while Snowflake and Raz had said goodbye before they headed out to Raz's Pawn and Antiques for the day. Over coffee, Snowflake had again bemoaned the fact that they had not found the Ark of the Covenant. Retta reminded her friend that there was more to life than his quest, and it was a blessing that they had all found their way out of the tomb unharmed. Snowflake had just shrugged and said that he was very excited that at least they did find the Seven Monks' Tomb.

Everyone was perplexed at who exactly Brother Eskinder had been, but Snowflake had nothing bad to say about the man and was just glad the priest, or ark-protecting-monk, or whatever he was had let them into the catacombs. They had all come to the logical conclusion, though, that Brother Eskinder's revealing smile proved the fact that the Seven Monks fraternity and protective order lived on. Surely, the ark had been moved by them to a new secret location.

The Sassafras twins had enjoyed the treasure hunt, but they were even more excited about all they had learned about the skeletal system. Both were brimming with optimism this morning because it looked like anatomy was going to be every bit as fun and exciting as zoology had been.

"Well, bro," Tracey declared, sitting back up. "I guess it is LINLOC time, right?"

"Right," Blaine answered, pulling out his smartphone.

The two had already sent all the SCIDAT information and pictures in to their uncle. Now, they would open up the LINLOC app to see where they were headed next.

"No way!" exclaimed Blaine, smiling. "We're going back to Australia! Well...as in 'we,' I mean me. This will be your first time."

Tracey smiled. This indeed would be her first time, and she was excited about it. They had gotten separated last time Blaine went to the land down under. She had zipped off to China on her own instead. That had happened when the Man with No Eyebrows had sabotaged their SCIDAT data, but this time Tracey hoped they could make it there together.

"The longitude is 151° 12' E and latitude coordinate is 33° 51' S. And it is in Sydney, Australia, to be exact," Tracey stated. "We are going to be gathering data on the respiratory system. We need info on breathing, the trachea, the larynx, the lungs, and alveoli."

"And our local expert's name is Julie Ette," finished Blaine.

The twins put on their harnesses and helmets and calibrated their three-ringed carabiners to the correct coordinates. As they let them snap shut, the special carabiners automatically clipped into the correct invisible zip lines, the lines that would take them to Sydney. The twins hung suspended in mid-air for a few seconds before they blasted off. It was now late morning, and the coffee shop was empty. Both twins wondered what someone would think if they walked in and saw two children floating in mid-air and then disappearing. They probably wouldn't believe their eyes, but, just as Uncle Cecil had instructed them, they'd been careful to keep the lines a secret. They always made sure that when they zipped off, they were alone.

Whoosh! The twins felt the lines take them. Their bodies were immediately enveloped in swirls of light as they zipped from Ethiopia to Australia. They had entered their SCIDAT data for the skeletal system correctly, so they were able to progress to the next location. For that they were glad because this was literally a rush!

Their bodies came to a sudden stop as the zip lining concluded. The twins were flooded over by a tingling sensation, momentary loss of strength, and temporary blindness. After all of their landings, they'd gotten used to these physical reactions to the light-speed travels. Now they just remained still, waited for their senses to normalize, and then they would find out where they had landed.

Polyester? Wool? Cotton? Silk? Blaine was no tailor, but he was pretty sure they'd just landed in a big pile of fabric, and were these feathers that were now draped across his face? The Sassafras boy began squirming around a bit, trying to figure out exactly where he was. He was just about to call out to Tracey when he heard a door crash open, and someone began yelling.

"Do you really think you can beat me? Don't you know who I am? I am Suzy McSnazz, seventeen-year-old pop sensation, and I am totally unbeatable! Those judges out there love me. The fans love me. Even the cameras love me. So don't you even begin to think that this lucky little run you have going will get you past me. I will destroy you in this round. I can't believe you even made it this far, all the way to the semifinals of TOBA. You're only fifteen and your songs are boring. It's inconceivable, but now that you are here, I am going to take you down, like all the other vocalists I've beaten so far."

"I wasn't saying I could beat you, Suzy," a calmer and quieter voice responded. "I was just talking about ventilation."

"You didn't say ventilation," Suzy McSnazz shouted back. "You said annihilation. You said you would annihilate me!"

"No, I was talking about ventilation. It's just another way of saying 'breathing'."

"Breathing?"

"Yes, breathing. Breathing is moving air in and out of the lungs. The process takes in oxygen for the body to use and removes the carbon dioxide from the body as waste. Oxygen is a type of gas found in the air that is essential to life. The body uses it to release energy from food. The body cannot survive without oxygen. Carbon dioxide is also a type of gas found in the air, but the body releases it as a waste product from making energy."

THE SASSAFRAS SCIENCE ADVENTURES

"I know what breathilation and oxinide and varbon dioxygen are," snapped Suzy. "You need to worry less about your fancy science and more about packing your bags, sweetie, because you are going home tonight after we face off!"

Blaine then heard some angry footsteps and another crashing of the door. He lay still for a few moments, and then he thought he could hear the sound of sniffling. Evidently, the quieter and calmer girl was still there. Blaine slowly pulled himself up out of the pile of fabric. When he managed to sit all the way up, he saw that Tracey, too, was just wiggling free from the material.

It looked now like they had landed in a big buggy full of costumes. That would explain all the different kinds of material and fabric. Both twins looked over and now saw the girl that had been left alone in the room. Her back was toward them, and she was sitting in front of a large mirror outlined in yellow light bulbs. The name "Julie Ette" was written on a sign above the mirror. So this was their local expert, and this must be her dressing room. The twins carefully and quietly climbed out of the costume-filled buggy and gently walked over toward Julie Ette. As the Sassafrases got closer, they could see that she was indeed crying. When Julie Ette saw them approaching, she quickly wiped her tears away and turned around to smile.

"Oh, hello. I thought I was alone."

"So sorry if we bothered you," Tracey apologized. "We were just going through the costumes."

Tracey looked over at Blaine, expecting him to add a comment to support her, but he was just standing there, speechless, with the silliest look on his face she'd ever seen. Apparently, he was also unaware that he had some kind of sparkly feather hat on his head. It must have attached itself to him when he was climbing out of the buggy. Tracey looked back at Julie Ette and laughed, a little embarrassed on behalf of her brother.

Julie Ette was strikingly beautiful, with brown eyes and

long, dark, wavy hair. "Oh, now I see," Tracey thought to herself. Blaine had what her father called, "the googly eyes." She slapped her brother on the back, trying to shake him from his love trance. It didn't seem to work.

"Are you okay?" Tracey asked, turning her attention back to Julie.

"Yes, I'm fine," Julie answered. "I was just trying to be nice to Suzy, but no matter what I do or say, she only sees me as competition. I was just trying to talk with her about proper breathing and how important it is for singers like us. Breathing is dependent on the diaphragm and the intercostal muscles between the ribs, which move to help air to be sucked in or pushed out. When the diaphragm moves down and the ribs move out, air is sucked into the lungs. When the diaphragm moves up and the ribs come back in, air is pushed out of the lungs.

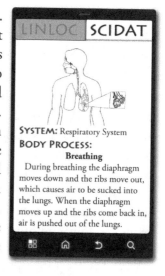

SYSTEM: Respiratory System
BODY PROCESS:
Breathing
During breathing the diaphragm moves down and the ribs move out, which causes air to be sucked into the lungs. When the diaphragm moves up and the ribs come back in, air is pushed out of the lungs.

Also, when breathing, air comes in through either the nose or mouth; then it passes into the throat, through the windpipe, or trachea, and into the lungs."

Julie sighed. "I know we're in a competition, but I just wanted to share with her some of the science that I love. I thought could help her become an even more talented vocalist than she already is, but, as she usually does, she just got mad at me. She is so intimidating. I don't know if I can go out there tonight and compete against her in the semi-finals."

"Sure you can," Tracey reassured, trying to be encouraging. "And besides, my brother and I, we love science. We would love to hear everything you know about breathing and the respiratory system."

"Really?" Julie Ette asked, surprised.

Tracey nodded. Blaine just kept standing there, looking goofy.

"By the way, we are the Sassafrases. I'm Tracey, and as I mentioned, this is Blaine. And you are Julie Ette, right?"

Julie Ette nodded.

"It's so cool that you sing AND enjoy science," Tracey said enthusiastically.

Julie's discouragement disappeared and she smiled and laughed. "It's so refreshing to meet some nice people around here. Show biz seems to stress most people out. Plus, you guys seem to love science! That is so cool!" Julie paused, still smiling, and then asked, "Do you want to hear something a little gross about breathing?"

Tracey nodded yes.

"When you're resting, air normally passes through your nose, right?"

"Right," Tracey confirmed.

"Well, the nose is warm and moist and covered in mucus."

Tracey scrunched up her own nose at this information. It was a little gross, like Julie had said, but it was also interesting and a little funny.

"Mucus traps harmful germs, dust, and pollen that is in the air we breathe," Julie continued. "Dust and pollen are also trapped in the hair-like surface of the trachea. Gross, but cool, huh?"

"Definitely," Tracey laughed a little.

Just then there was a knock at the door.

"Come in," said Julie Ette kindly.

The door opened and in walked four teenagers in a single file line. They stepped in sync until they were in front of Julie and the twins, and then they turned in perfect unison to face the three. Two of them were boys and two were girls, but they all looked very much alike.

"Hello," the one on the end announced. "I'm Denver."

Then, down the line they went. "I'm Dexter." "I'm Denise." "I'm Delores."

Then, in unison they trumpeted, "And we are the Colorado Quadruplets."

"Otherwise known as Ladder Smash," Julie let the twins know. "They do really amazing dance routines using ladders."

"So, are you ready for tonight?" all four quadruplets asked Julie in unison.

"I don't know," Julie answered. "I can't believe I even made it to the semi-finals. It feels like such a big deal. The semi-finals of TOBA."

"Whaf dis TLUBA?" were the sounds that suddenly tumbled out of Blaine's mouth.

Everyone looked at the Sassafras boy, who still had the feathery hat on his head, with perplexed looks on their faces.

"What did he say?" the quadruplets asked in unison.

Tracey was embarrassed for her brother. Not only were the googly eyes making him look goofy, but he couldn't even talk correctly in front of Julie Ette. "He asked: What is TOBA?" she answered for him.

The quadruplets broke from speaking in unison and fell back into the 'down the line' mode as they addressed Blaine's question.

"TOBA stands for 'Take Our Breath Away'," Denver started.

"It's an internationally televised talent competition," responded Dexter.

"With three celebrity judges and a pool of talented

contestants," added Denise.

"The talent is sectioned off into four different categories," continued Delores.

Then, like dominoes falling, the Colorado quadruplets listed the categories. "Singing." "Dancing." "Magic." "Miscellaneous."

"Wow! That sounds amazing!" Tracey exclaimed.

"It has been pretty amazing," Julie Ette agreed, with a humble smile. "Tonight is the semi-finals. We have made it down to the final eight. After tonight, there will only be four acts left, one in each category of talent. Those four will make up the contestants for the season finale of Take Our Breath Away, which will be filmed live, tomorrow night, here at the famous Sydney Opera House."

This was still very exciting to Tracey. "So who is still left in the competition?" she asked.

The quadruplets started answering again. "In singing, we have Miss Julie Ette versus pop sensation, Suzy McSnazz," Delores announced, going first and mixing up the Colorado rotation a bit.

Then Denise followed with, "In dancing, we have streamer dancer and ballerina, Trudy Stiles versus Ladder Smash, comprised of the Colorado Quadruplets."

"In magic, we have ventriloquist Flip Pippen with his magic puppet, Zippy. He's against the mysterious disappearing act known only as The Dark Cape," continued Dexter.

"And, finally, in the miscellaneous category, we have sports comedian Bob Squats versus Cletus Magnolia, the King Crab Whisperer," finished Denver.

Blaine, like Tracey, was very excited about this talent competition, so he attempted to speak again. "Sodees flisoof arehop flunit?"

Everyone just stared blankly at him.

"Oh, great. Crash and burn", Blaine thought. "What is

wrong with me? My heart is racing, my palms are sweaty, and my tongue is swollen."

"Ha, ha," Tracey said, laughing nervously. "Let me translate for my dear brother. He asked, 'So these face-offs are happening tonight?'"

"Yes," Julie Ette confirmed. "It all starts in less than an hour."

"On that note, we should go stretch," the quadruplets said together. "Good luck to you, Julie."

They then turned and walked out of Julie Ette's dressing room door in a single file line, just like they'd come in. Julie turned to the twins and smiled.

"I have some great front row seats reserved. You two are more than welcome to sit with me while I wait to go on."

"Wow! Really? That would be great! Thank you, Julie," Tracey returned, sincerely grateful and excited.

Blaine just stood with a goofy grin on his face and a sparkly feather hat on his head.

The Voice Box and Windpipe

The Take Our Breath Away talent competition was a top notch production. The Sassafras twins now sat on the front row with Julie Ette, one of the show's stars, trying to drink it all in. There was one main stage where most of the competition would happen, but attached to that stage were several additional levels of smaller stages where the contestants could move around a bit if they wanted or needed the space. Then, behind the stage were two huge video screens that were even now flashing up images of the competition up to this point. Also, behind the stage, to the left of the two main screens, were three personal sized balconies or perches. They were seemingly floating about twenty feet up in the air, with a perfect view overlooking the stage. Julie had told the twins that this is where the three judges sat.

Then, adding to all of this, everything was covered and filled with versatile and exciting lights. There were lights shining from within the stage itself. There were spotlights focusing on certain places and other lights sweeping around the giant room. There were sparkling lights and fading lights and lights in almost every color that one could imagine.

Topping everything off, there was a crowd made up of thousands of fans, sitting here in the Sydney Opera House, roaring and cheering, shouting and whistling, as they anticipated the start of this exciting semi-final round of this talent competition. This was almost sensory overload for the twins, who had spent nearly the entire day before underground in the tunnels of the dark and dusty royal catacombs in Addis Ababa, Ethiopia, but they were enjoying themselves, and they, too, were looking forward to seeing the show.

Suddenly, the crowd quieted, and all the lights faded. Now the sound of TOBA's electrifying theme song could be heard building in the speakers as it was being played off stage by a live band. At just the right moment in the song, the lights exploded with color and a tall handsome man in a suit jumped out happily onto center stage.

"Hello, there, mates," greeted the native Australian in a clear baritone voice. "My name is Dean Bean Junior, and this is Take Our Breath Away!"

The theme song rocked to its highest note, lights danced, and the crowd erupted in applause. Blaine and Tracey joined in, clapping their hands and letting out some whoops and hollers.

"Tonight," Dean Bean continued, "we will witness the semi-finals. After battling their way through some very talented competition, our final eight will go to war again. We now only have two groups or individuals left in each talent section. Those four talent sections are singing, dancing, magic, and miscellaneous. Tonight, our prestigious judges will decide the winners in each of those categories."

Dean Bean Jr. paused as light started to slowly rise on the three judge perches. "And our breathtaking judges are the legendary South American vocalist, Victoria Valencia!"

The crowd cheered as a beautiful Argentinian woman in her sixties waved and sat down. Dean then pointed to the next perch.

"The ground-breaking American entertainer, Bobbie Mega!" Dean continued.

The crowd continued its cheering as a striking African American woman waved and took her seat.

"Last but not least, the esteemed talent connoisseur from Scotland, Miles Dockerty!"

An unsmiling, unwaving man with slick hair, glasses, and a sweater-vest took his seat in the final judge's perch. The crowd noise died down a bit at the introduction of this last judge. Evidently, he was not as popular as the first two.

"And now, ladies and gentle-mates, say it with me," Dean Bean Jr. announced loudly and then paused, waiting for the voices of the crowd to sync up with his.

"Take Our Breath Away!" everyone shouted together.

Colorful lights exploded, and the band continued with the show's theme music. The stage was set. Now the semi-finals were about to begin.

After a TV time-out for commercials, the first contestant took the stage. The Sassafras twins watched with great curiosity as Cletus Magnolia, the King Crab Whisperer, walked out to the center of the stage where the stage-hands had already set up his large aquarium full of menacing-looking crabs. Without saying a word, he stepped up on the provided stool and plunged his hand into the aquarium. Immediately, the dozen or so King Crabs went frantic. They either fearfully scurried away from his hand or violently tried to pinch it. He quickly pulled his hand out of the water and looked at the crowd with a raised eyebrow. He then pulled out a pair of goggles and put them on. Now, he slowly stuck his face down into the water. No one could hear what he was saying (he was under water), but he started talking to the crabs. Bubbles came out of his mouth as he spoke, and he kept his head submerged for nearly a minute. He pulled his head out, took the goggles off, and prepared to stick his hand back down deep into the tank.

Cletus, with water dripping from his face, made eye contact with the crowd that was collectively wondering what was about to happen. All at once, he plunged his hand back down into the water. But this time the crabs didn't move. Not a single one of them. Evidently, whatever Cletus Magnolia had said to the King Crabs had calmed them down. He snapped his hand underwater, and all the crabs proceeded to line up in a straight line. He snapped his hand again. The crabs then carefully climbed up on top of each other's backs, making a sort of crab pyramid. Cletus snapped once more. The crabs climbed back down and lined back up. Now the King Crab Whisperer began moving his hand around in the water like he was directing an orchestra. The crowd could clearly see bubbles coming from different crabs in rhythm—it was like they were singing a fine-tuned underwater ballad. Blaine and Tracey looked at each other and nodded. "Pretty impressive," they thought.

After another commercial break, the streamer dancing ballerina, Trudy Stiles, was up. The TOBA band began playing a pre-selected song, and Trudy began her routine, dancing elegantly around the stage. She spun and twirled, did some jumps, and walked perfectly on the ends of her toes. All the while, she was swinging a long red streamer gracefully through the air. Somehow, she managed to never let any part of the streamer touch the ground. She was even talented enough to spell out words with the floating streamer as she danced. It was going to be hard for the Colorado quadruplets to top this performance.

During the next TV time-out, Julie Ette leaned over toward the twins and told them, "The next act is 'The Dark Cape'. He is famous for making things disappear. He puts on a great show, but everyone is a little afraid of him. He wears all black when he performs—black gloves, a black mask, and a black cape — but he never takes his costume off, not even backstage. None of us have ever seen his face, and we have never heard him speak a word."

"Weird," thought the twins.

When the show started again, the Sassafrases saw just what she'd been talking about. As the band played a slow, deep drum beat, The Dark Cape walked out into the spotlight. He was a big man draped in black from head to toe, and his mask was actually more like a helmet covering every inch of his head. His unseen eyes were hidden somewhere behind a dark visor. The lights came up a little, revealing several items lined up on the stage, ranging in size from a rather small soda can to a large refrigerator.

The Dark Cape reached down and picked up the soda can. He held it up high in his right hand for all to see. He then grabbed his cape with his left hand and started pulling it up to cover the can. He concealed the can behind the cape for only a split second, and then he dramatically jerked the black cape back downward, revealing that his right hand was now empty and the soda can had somehow disappeared. The crowd gasped in awe.

The magician repeated this amazing feat several times over as he continued down the line, each item getting bigger, more cumbersome, and harder to make disappear. Finally, he reached the heavy, tall refrigerator. Obviously, he couldn't pick this item up. He pushed on it a little and opened and closed its doors, showing the audience that it was indeed a real refrigerator. He then stood directly in front of it and slowly began raising his arms up over his head, each gloved hand grasping a corner of his cape. The cape spread out like wings and hid the refrigerator behind the magician. Then, all at once, he pulled the cape and his arms down, and the huge refrigerator was gone. The crowd gasped again and then applauded.

Another commercial break ensued as the Sassafras twins sat there on the front row, amazed at what they had just witnessed.

When the Take Our Breath Away show went back on the air, the auditorium seemed to pep up a little bit. It was Suzy McSnazz's turn to showcase her talent. The star was very charismatic and confident, and she got the crowd involved in her song right from the beginning. She danced around with her blond hair flying every which way. Suzy sang with emotion and expression on her face, with a voice that was cool and edgy.

Both twins looked over to see what Julie Ette's reaction might be to her competitor's performance. She was smiling and seemed to be enjoying Suzy's song. Julie noticed the twins looking at her, and this seemed to spark an idea in her mind.

"Hey, do you two want to hear something about the respiratory system?"

Both Blaine and Tracey nodded yes.

"Air from the nose and mouth passes through the larynx and into the trachea," Julie said. The twins were barely able to hear her over the music. "Inside the larynx are two bands of muscles called vocal chords. They are open during breathing but pull together when you talk or sing. The air passes through the vocal chords causing them to vibrate. The vibrations are heard as sound. So, right now, what you are hearing is the sound of Suzy's vocal chord muscles vibrating. It sounds beautiful, doesn't it?"

The twins were not only amazed at the science that Julie Ette was now sharing but also at the fifteen-year-old girl's attitude. Up on stage right now was the girl that was trying to knock Julie out of this competition. The girl they'd heard belittle and talk meanly to Julie in her dressing room. And now, here Julie was enjoying Suzy's performance and complimenting her abilities.

"What a class act Julie is," thought Tracey

"She's the prettiest and most amazing girl I've ever met," thought Blaine.

"The shorter the vocal chords, the faster they vibrate and

the higher they sound," continued Julie. "A woman's vocal chords are short and vibrate about two hundred twenty times a second, while men's vocal chords are longer and vibrate about one hundred twenty times per second. Isn't that crazy? The larynx is also known as the voice box, and the trachea is also known as the windpipe. The entrance to the trachea is covered by the epiglottis, which prevents food from going into the lungs. Then there are C-shaped pieces of cartilage along the trachea tube which helps to strengthen it. It's also lined with mucus and small hairs."

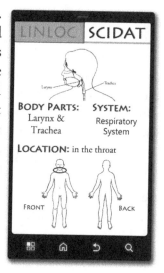

Tracey scrunched up her nose at the mention of mucus again.

Julie laughed. "Yeah, like I said earlier in the dressing room, the mucus traps dust, pollen, and other harmful particles away from the lungs, nose, and throat."

Blaine just smiled. Not even the mention of mucus coming from Julie could make him feel anything but enamored with her.

"Then, at the end, the trachea divides into two tubes, called bronchi," said Julie Ette finishing the information she knew about the larynx and trachea.

The three turned and watched the remainder of Suzy McSnazz's song. The pop sensation really was a great performer. The twins just hoped their new friend and local expert here in Sydney had a performance up her sleeve that could top the seventeen-year-old McSnazz. At the end of Suzy's song, the crowd cheered, the judges clapped, and Dean Bean Jr. came back out to the stage.

"Great job, Suzy McSnazz," said the host, as the blonde left the stage. "Ladies and gentle-mates, you have now seen the first half of tonight's acts. One performance from each section of talent, and it has been breathtaking, has it not?"

The audience roared in agreement.

"Well, don't go anywhere because we still have four acts left. And after these last four acts, our prestigious judges will decide right here tonight who will be moving on to tomorrow night's finals. So stay tuned!"

Lights danced, people clapped, and Take Our Breath Away cut to commercial. Julie Ette leaned over to talk to the twins.

"OK, Blaine and Tracey, it's time for me to go. I have to go back to my dressing room and get ready from my turn on stage. Wish me luck!"

"Good luck," said Tracey.

"Glud lik," mumbled Blaine.

Julie smiled as she left to go to her dressing room. Blaine just sat there dazed, watching her go. Tracey shook her head at her brother as she opened up the archive app on her phone. She started to look for pictures that would go with what Julie had just shared with them. She selected the best one and then closed the app, thinking that she could add the data later.

When the show started again, Bob Squats, the sports comedian was up. He was a heavy-set guy with a thick New York accent. He rumbled out on stage, grabbed his mic, and jumped right into his routine.

"So, the other day, I went to an all-you-can-eat-buffet with my buddy who is a NASCAR driver. I got my plate and was walking behind him, and I noticed he was holding his own plate at ten and two, and he was slowly weaving and swerving as he walked down the aisle. Suddenly, a waiter in front of him waved a green napkin, and he took off through the buffet like a shot out of a gun, but I noticed after a while that he was just going in counter-clockwise circles around the buffet bar, not stopping to get any food, though he did grab a jar of milk, I'm not sure why." Bob acted out all of his friend's movements as he spoke.

"Later, we were sitting at our table, and he told me he needed to make a 'pit stop,' so he got up to go to the bathroom, and, no kidding, he returned in fifteen seconds flat with his shoes shined, his beard shaved, and he said he was ready for another go-around at the buffet."

The crowd laughed, and Squats continued telling sports related jokes, most of which the twins didn't get at all, but they laughed along with the audience anyway.

After another break, it was time for Ladder Smash. The Colorado Quadruplets walked out onto the stage in their customary single-file line. They were wearing bright matching costumes that almost looked like gymnast tights, and they were all holding aluminum ladders. They stood still until their music started and then they burst into action. They swirled their ladders effortlessly over their heads. They climbed over, jumped through, and flipped off of their ladders in an amazing display of acrobatic talent. They threw the ladders back and forth to each other in a seamless juggling rotation.

Suddenly, the percussion in their song dropped out. In a planned twist to the act, the quadruplets took the ladders and started beating them on the ground to create the rhythm for the song. Occasionally, they would pick the ladders up and hit them together to make a different sound. They then used their hands to beat around on the rungs to make cooler percussion sounds than were in the song before. They finished their act by closing their ladders and slamming them down on the stage in perfect rhythmic progression. The crowd there in the Sydney Opera House cheered. It had been a great performance. The twins had no idea if the judges would choose them or Trudy Stiles for the finals.

After another TV time-out, the second-to-last performers came out to wow the audience—Flip Pippen and Zippy. Flip

Pippen the ventriloquist and his puppet friend, Zippy, were both dressed in tuxedos and had top hats on. Zippy did most of the talking and all of the magic; Flip just seemed to serve as more of an assistant. Most of the tricks he did were classics. He pulled a rabbit out of a hat, he linked two solid metal rings together, and he even made it look like he was sawn in two. But the most amazing thing about it was that Zippy was a puppet and he was doing all of this. The Sassafrases could not figure out how, either.

When the magic duo's act finished, the entertained crowd applauded. Now there was only one more act left—Miss Julie Ette. The twins could hardly wait.

This time, as TOBA cut to commercial, the Opera House cameras swept around the audience and displayed images of cheering fans up on the screens. Blaine and Tracey were surprised and excited to see their own faces flash up on the live video footage, but as the cameras continued to sweep around, the Sassafras twins saw something displayed on the screens that rocked them to the core.

They saw the face of the Man with No Eyebrows. He was here even now, somewhere in the sea of cheering fans.

CHAPTER 5: THE OPERA HOUSE

At Last, the Lungs

From the moment Julie Ette opened up her mouth to sing, everyone in the Sydney Opera House was completely still and silent, captivated by her voice. She sang slowly and with conviction. She let her notes float out beautifully into the air, like gifts for the ears to catch, take in, and then give to the soul to ponder. The band played the music to her enchanting song perfectly, but the music and even the song itself were secondary. It was her voice that told the story. It was the vibrations of her vocal chords that were leaving all in awe. Julie Ette had the most beautiful voice the Sassafras twins had ever heard.

Time seemed to stand still as their local expert sang. Blaine and Tracey momentarily forgot all about their sighting of the Man with No Eyebrows. They just sat there on the front row, completely and utterly enthralled. When Julie Ette finished her performance, everyone remained silent. There was no clapping or shouting because every person was blissfully frozen in wonder.

Julie put the microphone down and slowly walked off the stage, her light footsteps the only sound in the auditorium. A few moments later, Dean Bean Jr. had the difficult job of breaking the silence. He lifted his finger to strike the band back up and made the announcement that after one last commercial break, the judges would render their decisions on whom the finalists would be.

During the TV time-out, the crowd inside the Sydney Opera House, slowly but surely, began to break out of their silent trance.

Tracey pulled at her brother's arm. "Let's go backstage real quick and see if we can find Julie to congratulate her on such a great performance."

Blaine nodded and the two jumped up and headed to the back. But what the twins found when they arrived backstage was not at all what they had expected. It wasn't a group of nervous performers anticipating the impending results of the semi-finals. It was a cast in chaos. There was a thin pale man the twins didn't recognize tied up in the corner, and Suzy McSnazz had fallen into the King Crab tank. But she hadn't just fallen in. She was also trapped. It looked like an entire stack of crates had fallen over on top of the aquarium, blocking her in the tank underwater. Evidently, this had just happened because everyone was still scrambling around trying to figure out what to do.

"Hurry!" Julie Ette screamed. "We've got to get her out of there! The body can survive days without food and water, but only minutes without oxygen!"

Bob Squats, the Colorado Quadruplets, and Flip Pippen and Zippy all rushed over and started trying to hoist the heavy crates off of the aquarium. Cletus Magnolia quickly knelt down and started mumbling through the glass of the tank, urging his King Crabs not to pinch the pop sensation.

Meanwhile, Trudy Stiles went over and tried to console the man who was bound. He was very exasperated and seemed to be hyperventilating. The twins jumped in and tried to help move crates.

"Even though the lungs are the largest organ in the body, they still only have a limited amount of space for air. Suzy is running out of that air right now! We have to hurry!" Julie said while frantically pushing at a crate. The group was attempting to hoist away the crates from the top of the pile first.

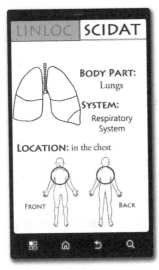

"The lungs sit on each side of the heart in the chest and are protected by the spine and rib cage. They rest on the diaphragm, which is a muscular sheet that separates the abdomen and the chest cavity. A thin membrane covers the lungs and another covers the inside of the chest. There is fluid between these two membranes, which decreases the friction and prevents pain during breathing. This is called the pleural cavity."

Julie Ette kept giving information as the big stack of heavy crates was being moved by the group of entertainers. The twins assumed this was calming Julie and keeping her mind off the fact that Suzy McSnazz was slowly drowning.

"The lungs consist of five lobes: two lobes in the left lung and three lobes in the right lung. The left side of the lung only contains two lobes so there is room for the heart. Each lobe sits in its own pleural cavity. The two sides of the lungs are connected by the bronchi and the trachea."

Suzy really began squirming around there in the tank.

The twins could see the look of panic on her face. Cletus, with his whispers, was somehow succeeding at keeping his king Crabs away from McSnazz. They were all piled up in the corners of the aquarium, blowing bubbles. The grunting, sweating, determined group of performers was getting close to freeing the pop singer. They just all hoped they could free her in time.

"The environment in the lungs is very moist," continued Julie, with worry in her voice. "Healthy lungs are pink because they are full of blood. They are spongy because they consist of a branching network of tubes, called bronchi and bronchioles, which end in air sacs called alveoli."

"Just three more crates!" shouted Bob Squats, the hefty comedian.

He, Flip Pippen, and Zippy got one. The Colorado Quadruplets moved another, and Julie Ette and the Sassafras twins hoisted off the last crate that was blocking Suzy McSnazz in the King Crab tank. The second the last crate was moved out of the way, the seventeen-year-old blond-headed singer burst out of the water and stumbled out of the aquarium, gasping for air. She fell to the ground, dripping wet, and started to breathe in deeply. She began to cry, and then she started to sneeze.

Julie immediately knelt down beside her and tried to comfort the girl. But Suzy McSnazz would have none of it. "Get …away… Achoo!...from...Achoo!...me!...Achoo!"

"Suzy! I'm just glad you're okay," exclaimed Julie Ette in a sincere voice. "I was scared we weren't going to get you out in time!"

"I told you to leave me alone," yelled McSnazz in between sneezes, actually pushing Julie Ette away.

"Whoa, whoa, whoa," Bob Squatts said as he stepped in. "Now hold on a minute here, Suzy McSneeze. Julie's just trying to help you out. She was the one that was urging us to hurry and get you outta there. What happened to you, anyway? How did you get trapped in the aquarium and under the crates?"

Instead of answering, Suzy just kept crying and sneezing.

"What's her deal?" asked Flip.

"Yeah, why all the sneezing?" added Zippy.

"During a sneeze, air rushes out of the lungs and burst through the nose," explained Julie. "Droplets with harmful bacteria, pollen, or dust are released into the air, but I'm not sure what is causing Suzy to sneeze so much."

"Hey guys!" Everyone was interrupted by the loud call of Trudy Stiles. "Can someone please grab that inhaler? I think that's this guy's problem. He needs it and he's pointing to it!"

All turned to see Trudy across the room assisting the thin pale man. She had untied his hands and feet, but he was still gasping for breath, and he was now pointing toward the inhaler on the floor.

Blaine spotted the inhaler first, and he quickly scooted over to get it and then give it to the man. The man reached out with a bony hand, grabbed the inhaler, put it up to his mouth, and then deeply inhaled the medicine. He repeated these last two steps a couple more times, and within seconds he was breathing much easier. Everyone just kind of looked at him, not sure exactly what to say. They were all wondering who he was until he spoke.

"I can tell you how Miss McSnazz over there got trapped in the tank," he said, weakly. "And I can tell you what happened to me, too."

The group gathered in to better hear his low and shaky voice.

"First, let me say I'm sorry that you all had to see me and meet me like this. My name is Phil Earp, and I am the Dark Cape."

"The Dark Cape?" everyone gasped together.

"But that can't be!" Trudy Stiles exclaimed. "The Dark Cape is a much bigger man."

"Yes, I know," acknowledged Phil. "I had the suit especially designed to expand around my small frame, making me look big and imposing. But it was truly me that was inside the suit. That is, until the suit was stolen."

"Stolen?!" the whole group cried out again in unison, alarmed.

"Yes, stolen," he answered, as he started wheezing again. He took a few more puffs from his inhaler and then continued. "After my act, I came backstage to rest and watch the rest of your acts on this monitor." He pointed up to a television screen anchored in the wall just above them. "I was enjoying your breathtaking song, Julie, when suddenly I was hit hard in the back and knocked to the ground. I looked up to see a strange man with no eyebrows attacking me. He ripped the Dark Cape suit off of me, tied me up, and then put the suit on himself. Just as he was running to leave, Suzy McSnazz came around the corner, and he crashed right into her. She fell into the crab aquarium, and then he started going bonkers. He must have accidentally pulled the vanish string."

"The vanish string?" the amazed group exclaimed. They were still somehow answering in unison.

"Yes, the vanish string," Phil Earp answered. "There is no way I could do my magic without that suit. As all of you know, my specialty is making things disappear. I have made all kinds of things disappear throughout this TOBA competition, but I was saving the best for last. I developed a 'vanish string' and had recently added it to the inside of the Dark Cape suit. If I made it to the finals, I was going to pull this string and make myself disappear."

"Yourself?" everyone asked.

"Yes, myself," Phil confirmed. "So this, I believe, is why the man that stole my suit was going haywire. He was probably unaware that he was tugging on the vanish string repeatedly, making himself disappear and re-appear in quick and abrupt intervals. It can be quite shocking to make yourself disappear, and he probably wasn't expecting it. He was stumbling around rather violently as he was appearing and re-appearing. That is when he crashed into that stack of crates, knocking them over on top of the aquarium, and blocking Suzy McSnazz underneath the water. I wanted to help her, but I was tied up, and on top of that I started hyperventilating because of my asthma."

"Asthma?!" the group asked.

"Yes. Asthma," Phil answered, while taking another puff of medicine.

"Asthma is a common disorder that causes the tubes in the lungs to swell. This decreases the space for air to move through, making it difficult to breathe," Julie Ette explained. "The inhaler Phil's using has medicine in it that helps open up the airways, allowing him to catch his breath and breathe easier."

"Well, we gotta catch this guy," Bob Squatts said, on the verge of anger. "We can't let a thief come in here and get away with Phil's Dark Cape suit!"

Everyone in the group agreed, but before they could formulate any kind of plan, they were interrupted by a frantic Dean Bean Jr.

"What are you still doing backstage, mates?" His voice was not calm but flustered. "The judges have made their decisions! All of you must come back out to the stage right now!"

A group of stagehands rushed in behind Dean Bean, and corralled the performers, and then ushered them back out to the stage. Blaine and Tracey quickly left the backstage area by a different route and returned to their seats in the front row. Dean Bean Jr. soon appeared in the spotlight at the center of the stage. His customary smile was now back on his face, with no trace of angst.

"Ladies and gentle-mates!" he said with his clear baritone voice, "we have finally arrived at the moment you've all been anticipating! The judges have made their decisions! We will now find out which performers will go through to the finals of Take Our Breath Away!"

The always-lively crowd cheered at the host's announcement. Lights came up on the three judges' perches. Victoria Valencia and Bobbie Mega both smiled happily and waved to the audience. Miles Dockerty just stood there with his arms folded, looking perturbed or possibly bored. Then, as the stage lights started to come up, the eight acts of talent walked out and lined up for all to see.

"OK, performers," said Dean. "When I call your name, please step forward."

The host paused as the band began playing a heart-pounding song. The lights swooped down in dramatic fashion.

"In the miscellaneous category, we have Cletus Magnolia and Bob Squats." The King Crab Whisperer and the comedian each stepped forward as their names were called.

"The miscellaneous artist going to the Take Our Breath Away finals is..." The percussionist in the band started a drum roll as Dean Bean left everyone in suspense for a moment.

"...Bob Squats!" the host shouted in elation, pointing at the sports comedian.

The audience applauded. Bob smiled, bowed and mouthed big 'thank you' to the judges. He then hugged his disappointed but gracious competitor, Cletus.

Dean continued on, "Trudy Stiles and Ladder Smash, please step forward to represent the dancing category of talent."

The Colorado quadruplets stepped forward in their signature cohesive style. The streamer-dancing ballerina stepped forward as well.

"The dancer or dancers moving on to the finals of Take Our Breath Away is..." There were more drum rolls for suspense. "... Ladder Smash!"

Again the crowd cheered at Dean's announcement. The quadruplets gave each other energetic high-fives, waved their appreciation to the judges and fans, and embraced Trudy, who was sad for herself but excited for her fellow dancers.

"In magic," Dean announced next, "our semi-finalists are The Dark Cape and Flip Pippen and Zippy!"

The ventriloquist and his magic puppet stepped forward from the line and smiled, but the pale and thin Phil Earp remained stationary, seemingly frozen stiff. He was so used to being in his dark cape suit that he must now have felt open and exposed. Dean Bean Jr. walked over, put an arm over Phil's shoulder, and gingerly pulled him forward.

They were perplexed murmurings coming from the audience. There were obviously confused. They had no idea who this little white man was that Dean was standing next to. They were all expecting to see the Dark Cape in his usual costume.

Bean Jr. continued, in spite of all the confusion. "The magician moving forward to the Take Our Breath Away finals is ... the Dark Cape!"

Some in the crowd clapped, but most were still looking around for the Dark Cape that they recognized. Flip and Zippy walked over to congratulate Phil.

"Well done, Phil," congratulated Flip.

"And don't worry, friend," Zippy added. "We will get that suit back before the finals."

"And last but not least!" Dean Bean announced into the microphone, "Suzy McSnazz and Julie Ette, please step forward."

Julie took a step up, smiling but nervous about the impending decision. Suzy also stepped forward, but she was not smiling in the least. She was still dripping wet from her tumble into the King Crab tank. Her blond hair was stringy and soaked. Her make-up was blotched, and her nose was running from all the sneezing she'd done earlier. Now, for the second announcement in a row, the confused crowd had no idea what was going on.

One could almost hear their collective question. "What happened to Suzy McSnazz, and where is the Dark Cape?"

The Sassafras twins were the only audience members that truly knew what had happened backstage.

"These two talented young women are our semi-finalists in the singing category," Dean said. "And the vocalist that has made it through to the season finale of Take Our Breath Away is …"

The Alveoli Adventures

He paused and let the drama build and the drum roll linger. Blaine and Tracey held their breath, hoping with all their hearts that their local expert's name was about to be called.

"Miss Julie Ette!" Dean Bean announced jubilantly.

Julie covered her open mouth with her hand in sincere surprise. She had done it! She had made it to the finals of this prestigious talent show. Suzy McSnazz, however, was crushed. She slumped down on the stage in a heap of sobs, crab water, and runny mascara. Dean Bean knelt down and tried to help Suzy get up, but the defeated vocalist just screamed at him in front of the crowd and the international audience.

"Get away from me, Bean!" She pushed the host away and then pointed at Julie Ette. "She wasn't supposed to beat me. I'm Suzy McSnazz, seventeen-year-old pop sensation! I was supposed to be the Take Our Breath Away champion! Me, and nobody else!"

McSnazz then rolled over a couple of times and literally crawled off the stage. Dean Bean Jr. watched her in disbelief, but then he turned back towards the audience with a smile on his face.

"Well then, ladies and gentle-mates, there you have it. Tonight, four were eliminated, but you now have your Take Our Breath Away finalists! And here they are: Bob Squats, Ladder

Smash, The Dark Cape, and Julie Ette!"

The lights exploded again in dazzling brilliance, and the TOBA band energetically played the show's theme song. The Sassafras twins looked at each other and smiled. They were satisfied with the judges' decisions.

That night Blaine and Tracey were able to sleep on a couple of cots that were in the vacated dressing room of Suzy McSnazz, but before they went to sleep, they entered all the SCIDAT data they had gotten so far on this leg of their journeys.

They woke up the next morning thinking about the Man with No Eyebrows. It was no longer a question of "if" he would show up. It was now a question of "when" and "how," but they still didn't know "why." That was the burning question. Why was he so determined to mess things up for them? What reasons did he have for wanting to stop a couple kids from learning science?

Julie had invited the twins to meet her that morning in a place called the green room. So Blaine and Tracey headed that way. When they entered the green room, they saw Julie, Bob, Phil, and the Colorado four all standing around a table that was stacked high with food and drinks. The seven performers waved Blaine and Tracey over to join them.

"I had some pretty amazing stuff planned for tonight," the twins heard Phil Earp say, as they walked up to the table and each grabbed a bagel and a boxed milk. "But without that suit, the magic is as good as gone. I guess it will be more like the final three tonight than the final four."

"Oh, c'mon, guys," Bob remarked, "don't talk like that. We are gonna find your suit. I mean, I want to beat you in this competition and all, but I want to beat you fair and square. You know, man to man ... or man to cape."

Suddenly, appearing out of nowhere, there he was. The Dark Cape, standing in a far corner of the green room! The nine who were eating their breakfast were shocked. They had just been talking about him, and now here he was, staring right at them. Blaine managed to get something out of his mouth first, but sadly, because of Julie's presence, it was still unintelligible.

"Duh nanwano oobrawds! Utesga hum!"

Now everyone was shocked and confused.

"What did he say?" asked Bob.

"The Man with No Eyebrows," Tracey translated. "Let's get him!"

"Yes! Let's get him," Bob agreed. "Let's get Phil's suit back!"

The seven performers started in the direction of the black caped thief. They couldn't see his face because it was covered by the black helmet and visor, but he was visibly flustered. He was obviously looking around for a way to escape. Evidently, the Man with No Eyebrows still hadn't figured out how exactly Phil's suit worked. He took a quick step to his left and then disappeared again.

"Wha...where did he go?" Bob asked.

His question was answered immediately as the Man with No Eyebrows now appeared on the other side of the room. The group ran as one, now to that side of the room. Right before they reached him, he disappeared again. The group stumbled to a stop. This sequence of events happened over and over again throughout the green room, until finally Phil managed to reach out and grab the edge of the cape while the man was visible. The Man with No Eyebrows jerked himself away to escape, but he ended up crashing into the food and beverage table.

The group of nine lunged at him in an attempt at a dog-pile tackle, but he disappeared just in time, causing the group to tackle not him, but instead to tackle armfuls of bagels, bananas, cereal, juice, milk, and yogurt. They picked themselves up, and they were still determined as ever to catch this eyebrow-less thief. There he was again, over by the door, pulling it open to leave!

THE SASSAFRAS SCIENCE ADVENTURES

"He's getting away!" Bob called out.

The Man with No Eyebrows jetted out of the door into the backstage area. His nine pursuers quickly followed. They had resolved to catch him, but it was proving to be very difficult to even keep an eye on him, much less actually catch him. Basically, they were chasing a teleporter. He ran through the backstage area, appearing and disappearing as he went, and then out onto the stage.

The stage was empty and silent right now, and the entire auditorium was vacant and dark. The Man with No Eyebrows ran across the stage and jumped off the other side, landing in front of a row of seats. He quickly rushed around to the end of the row and raced up the aisle toward the back of the auditorium, disappearing into the shadows.

The group of nine paused for a moment in their current spot on the stage, straining their eyes to catch their next glimpse of the disappearing man. Suddenly, a shot of light shone through the darkness as they all spotted the Dark Cape leaving through an exit in the rear of the auditorium. They all jumped off the stage and ran up the aisle in pursuit.

Denver reached the exit door first, then Dexter, then Denise, then Delores. The Quadruplets were quickly followed by the other five. They now found themselves in the Opera House's lobby.

"Look, there he is. Outside," shouted Tracey.

The Man with No Eyebrows could now be seen running in the black caped suit outside on a walkway that ran in between the building and the harbor. With no hesitation, the group ran outside in chase. He disappeared again, then appeared and then disappeared once more. It was truly exhausting, running after this guy. They ran around and around the Sydney Opera House chasing the Dark Cape, but eventually, he permanently eluded them.

This became apparent when they stumbled upon a heap of black cloth. Evidently, the Man with No Eyebrows had had enough of the Dark Cape suit, so he'd taken it off and left it in a pile before he had fled. The man was now gone, but at least Phil Earp had his suit back. Everyone congratulated the skinny pale magician on having his Dark Cape suit back.

They all joyfully headed back inside to spend the day preparing for their final performances. They knew that tonight was the night that Take Our Breath Away would crown a champion.

Blaine and Tracey Sassafras spent most of the day resting, which they figured their bodies really needed after all the zip line traveling, sonic-lag, and crazy adventures. Later that evening, just a few minutes before the Take Our Breath Away final show was about to start, they stopped by Julie Ette's dressing room to wish her luck. They liked all the other performers who were left and had become friends with them too. Even so, Julie was their favorite, and they really hoped that she would win this thing.

Tracey knocked, and they heard Julie say, "Come in." The twins entered and saw that Julie was sitting at her seat in front of the yellow-bulb-lined mirror.

"Hey, guys!" she said cheerfully but nervously. "I was just going over a little more of the science that I know about breathing and the respiratory system. Science just seems to calm me down when I'm nervous. Man, am I nervous right now! The Take Our Breath Away finale, I can't believe I actually made it!"

"Leekan bleeve noddit," Blaine blurted out, googly-eyed.

Tracey just shook her head. Her poor brother still hadn't gotten out a successful sentence in Julie Ette's presence.

"He said, 'We can believe you made it'," Tracey explained. "And we can. You are a gifted singer. Just go out there and do what you are so good at doing—singing! You are guaranteed to take our breath away."

"Thank you, you two," Julie said sincerely.

"What science were you going over?" asked Tracey.

THE SASSAFRAS SCIENCE ADVENTURES

"The alveoli," the vocalist replied.

"The alveoli?" asked the Sassafras girl, not at all sure what that word meant.

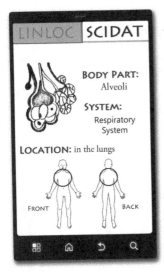

"Yes, the alveoli," Julie confirmed. "They are the place in your lungs where gas exchange takes place. Alveoli are tiny sacs or bags that are found in the lungs at the very end of the air tubes. They are always in clusters, and they resemble a bunch of grapes. There are over four hundred million alveoli in your lungs and each one is covered in blood vessels. These vessels deliver the body's waste carbon dioxide to the lungs and take away the life-giving oxygen. The walls of the alveoli are thin, which allows for easy exchange of these gases. Get this, the combined gas exchange space that the alveoli cover is equal to the size of a tennis court. Isn't that amazing?"

"Wow, that is amazing," replied Tracey, genuinely impressed.

"Just the respiratory system as a whole is amazing!" Julie continued, "The system of tubes that carry air in and out of the body consists of the nasal cavity, throat, larynx, trachea, and lungs. If you wanted to, you could sort of include the ribs because they surround and protect the respiratory system as well as play a role in breathing, but really, the ribs are a part of the skeletal system."

Just then, there was another knock at the door as Dean Bean Jr. peeked his head into the room.

"Julie Ette," he declared, "it's time. You're up first tonight. Are you ready?"

CHAPTER 6: FINE DINING IN ITALY

Use-a Your Brain

"Are you ready?" Cecil Sassafras said to Aristotle, one of his plastic skeletons. "The twins just sent in all their data on the respiratory system, so now it is time to add all of this to you!"

Cecil was holding a cardboard box full of different model parts to a plastic respiratory system set. Blaine and Tracey's uncle hummed a happy song to himself as he set the box down. Then he started carefully snapping the different pieces in place, mainly inside Aristotle's rib cage.

Unbeknownst to the twins, as they were zipping around the world gathering anatomy data, their uncle was also adding plastic pieces to his two skeletons: Socrates and Aristotle. Cecil was hoping to surprise the twins with the revelation of these two guys when the children returned after completing everything required of them on their anatomy leg of their summer travels. Right now, Cecil Sassafras was giddy with joy. Everything that he and President Lincoln had planned out with the zip lines was working perfectly. Blaine and Tracey were doing even better than he had expected. Even more than that, the twins weren't just learning science. They were loving it.

Ladder Smash had given an even better final performance than the one the twins had seen in the semi-finals. This time they had stacked their ladders on end, four high, and had created cool sounds while doing acrobatics and dancing up and down the precarious tower of balancing ladders. Bob Squatts had kept the crowd in a continual uproar of laughter throughout his comedy routine as he talked about 'the practice swing' in golf, the 'flail' in soccer, and other sports themed jokes.

Finally, Phil Earp gave a truly wondrous performance as the Dark Cape. As was his custom, he had lined up several items on the stage in order of size. He went all the way from making a key disappear to making a car vanish. Then, as he had promised, he had made himself disappear. The crowd had gasped in wonder. Even though Blaine and Tracey had knew about the 'vanish string,' they were amazed and surprised.

All three of these performances had been championship caliber acts, but even so, they had all paled in comparison to the first performance of the night. Julie Ette's enchanting song had again taken everyone's breath away. After all, thought the twins, that was the name of the show, was it not? Julie Ette had cried tears off happiness as Dean Bean Jr. had announced her as the winner. The judges had selected her. The joyous crowd had applauded her, and she had been showered with balloons and confetti falling from somewhere up in the ceiling of the Sydney Opera House.

Blaine and Tracey had been thrilled that their local expert had won. Backstage, they had joined the line to congratulate her behind Bob, Phil, and the quadruplets. She had done it—she had won Take Our Breath Away. After Tracey had her chance to congratulate Julie, and Blaine had his chance to say, "coin gat sons," the twins had retired to a hidden spot to finish putting in their SCIDAT data and sending in their last picture.

It all went through, and now LINLOC was telling them that their next location was Venice, Italy: Longitude 12° 34' E, Latitude 45° 43' N. Their local expert's name was Vittorio Benaneli, and they would be gathering information on the brain, the spinal cord, and the five senses.

"Ahhh, Venice," Blaine sighed, as he and his sister calibrated their carabiners and clipped into the next invisible zip line.

"I just hope you can talk correctly there," Tracey quipped. Both twins chuckled as they were swept off at the speed of light. The first thing the Sassafrases realized when they came to a stop was that they were cold, very cold.

"I didn't know Venice would be this cold," shivered Blaine. "It feels like we are back in the arctic."

"Yeah, it's f-f-freezing," Tracey agreed.

As soon as their strength returned, the twins started taking their harnesses and helmets and began to move around to warm up a bit. As their sight returned, they saw that they'd landed, not in the arctic, but in a walk-in freezer. There were shelves on either side of them, stocked full of boxes and bags of what looked like mostly frozen meat. Blaine went over and tapped a thermometer on the wall. There was hardly any red visible at all.

"I think it's c-c-colder in here than it was in Alaska," the Sassafras boy stated. "C'mon, let's g-g-get out of here."

He grabbed the handle of the one and only door and pushed, but nothing happened. The door did not open.

"Here, let me try," Tracey gasped, pushing her brother aside. She pulled, pushed, twisted, and cranked at the frozen handle. She reaped the same result as Blaine—the door would not budge.

"Well, great," Tracey sighed, frustrated. "We are locked inside a d-d-deep freezer. What are we going to do now?"

Blaine shrugged. "I don't know."

The twins just stood there shivering, not sure exactly what the best course of action was. They were slowly but surely turning into popsicles.

"I guess we c-c-could c-c-call Uncle Cecil and see if he knows why we l-l-landed in here," Tracey suggested.

Blaine nodded like he thought that might be a good idea. Just then, the door to the freezer was yanked open from the other side. Now, in the open doorway in front of them, stood the biggest, roundest chef the Sassafrases had ever seen. He was dressed in white, complete with an apron and a tall chef's hat. He had a thick black mustache and rosy cheeks, but his cheeks weren't rosy from being merry. They were rosy from being angry, and the more Blaine and Tracey stood speechless, the redder those big cheeks got.

"What in-a the name of-a Luigi are you two doing in-a the freezer?!" the big chef asked in a loud booming voice, with a rhythmic bouncy Italian accent. "You two look-a like a couple of gelato pops! You must-a forgotten rule-a number one fifty three, 'Don't-a let yourself get-a locked in the meat locker!'"

Blaine and Tracey both just continued to stand there, frozen—both from the chill of the freezer and from fear of this man.

They had no idea what to do or say. Behind the big chef, the twins could see a small crowd of additional chefs gathering to see at whom the big guy was yelling.

"You two need to use-a your brains!" the big chef shouted while shaking his outstretched arms, shoulders shrugged, elbows in, like he was completely flabbergasted at the twins' incompetence.

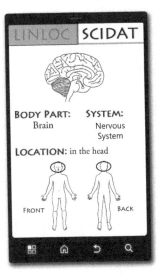

BODY PART: SYSTEM:
Brain   Nervous
              System
LOCATION: in the head

FRONT   BACK

The big man then quickly swiveled around and addressed all the gawking chefs now in the kitchen. "You all need-a to learn how to use-a your brains. Do I have-a to do everything around here-a myself? Do you not-a know that-a the brain is the control center for the body? It is-a where you think, feel, and-a move. The brain is also responsible-a for your memory. It-a contains over one hundred billion neurons, and happens to look-a like a mushy gray sponge-a. Folds and-a grooves in-a the brain increase the surface area, allowing the brain to fit inside-a the skull. The brain is-a constantly receiving and-a processing information, even when we are asleep-a."

Many of the chefs were rolling their eyes like they had heard this brain information a thousand times, but some looked like they were just as scared as the Sassafrases were. The big chef, who the twins were now sure was Vittorio Benaneli, their new local expert, continued.

"The left half of-a the brain controls the right side of-a the body, and the right half of-a the brain controls the left side of-a the body. The brain consists of-a three main regions: the brain stem, the cerebellum, and the cerebrum. The brain stem controls functions essential to life-a, such as-a breathing. The cerebellum coordinates balance, posture, and-a movement. The cerebrum is-a the feeling and-a the thinking part of-a the brain. Its-a divided into two halves connected by-a the corpus callosum. The front of-a the cerebrum is-a responsible for controlling thoughts, emotions, learning, and-a consciousness. While all other parts of-a the cerebrum are responsible for controlling speech, fine-movement, touch, hearing, sight, smell, and-a taste."

Vittorio turned back, grabbed the twins by the collars of their shirts, and pulled them out of the walk-in freezer. He closed the door behind them and then continued, "There are-a two types of actions in-a the body: voluntary and involuntary. Voluntary are the actions that-a your brain controls consciously, such as-a kicking a ball-a. Involuntary are those actions that-a your brain does-a not control consciously, such as-a breathing or digestion. The brain is-a the control center for-a the nervous system! The nervous system controls-a the body through a communications network. Messages are carried through-a the body much-a like electricity is carried by power lines. The nervous system has-a two parts: the central nervous system, which-a includes the brain and-a the spinal cord, and-a the peripheral nervous system, which-a is made up of a network of cells called neurons."

The big chef's cheeks were as red as red could be, and he was beading with sweat from his speech. All along, the twins had thought Vittorio had been mad throughout his rant and information giving, but as he now eyed everyone in the kitchen, a slight smile formed on his face, but it was gone as fast as it had appeared.

"By the looks of-a your faces," Vittorio bellowed, "you have-a all forgotten rule-a number forty-seven: Know everything you can about-a the brain then use-a it!"

Vittorio reached up and twirled the ends of his big mustache with his fingers. "Now, what is-a everyone just-a standing around for?! Get-a back to work!" He looked at Blaine and Tracey and pointed. "You two, get over there and-a help Giovanni with-a the dishes!"

The big bouncy chef then exited the kitchen through a pair of double swinging doors. Immediately, the entire kitchen staff jumped back into action, manning their cooking stations, firing up stoves, shouting out orders, chopping vegetables, mixing ingredients, and tasking themselves in what looked like organized chaos. The twins slowly and carefully made their way through the hustling crowd of chefs to the back of the kitchen, where they found stacks and stacks of pots, pans, and assorted dishes piled high on a stainless steel counter.

A short Italian boy about their age was quickly shuffling the mountain of dirty dishes around, trying to keep it from falling. With a big smile on his face, he turned around and introduced himself.

"Hello-a my friends! My name is Giovanni, and I could definitely use-a some help."

Blaine stepped in to steady a wobbly stack of bowls that looked like it was about to fall. "Nice to meet you, Giovanni," Blaine responded. "I'm Blaine and this is my sister, Tracey, and we would love to help."

"Grazi!" Giovanni exclaimed. "Grazi, grazi, grazi! One of you can-a wash, one of you can-a rinse, and I will put the cleaned dishes on the drying rack. How does that sound-a?"

"Sounds great," the twins said in unison.

Blaine began to rinse, and Tracey began to wash. Giovanni started the all-important task of stacking the washed and dried dishes on the drying rack. He was quite skilled at this task, too. The twins watched in wonder as he took the washed dishes flipped and spun them, and then placed them on the drying rack without breaking a single one.

"So, is Chef Benaneli angry all the time?" Blaine asked Giovanni as they worked.

"Uncle Vittorio?" Giovanni asked. "Oh, no, he was not-a

angry tonight. That was-a just passion. I want to be just like him when I am a man. He is-a the best chef in all of Venice, and he is also the smartest man I have ever known."

"Chef Benaneli is your uncle?" Tracey questioned.

"Oh, yes-a! Vittorio Benaneli is my uncle," Giovanni beamed with a smile. "Do you not see the family resemblance?"

The twins looked over at the Italian boy and saw that he looked just like a miniature Vittorio. They nodded their heads 'yes' to Giovanni's question. The boy smiled and continued flipping around dishes on the drying rack.

"My appearance is similar to my uncle's, but most of all I hope that my passion in the kitchen resembles his," he shared. "Uncle Vittorio gave me this job last year when I turned-a thirteen. I am so grateful, but I don't want-a be a dishwasher for the rest of my life, you know. I have-a been working on my cooking skills in my own-a free time, and I hope-a to become a pastry chef soon. Then I want-a to get promoted to sous chef. After that I want-a to eventually be the executive chef of my own restaurant. Just-a like Uncle Vittorio. That is just what he did. He started as a dishwasher, and now he is-a the owner and executive chef of this, the best-a restaurant in Venice: Benanelis!"

Just then, the three dishwashers heard the huge crash of shattering dishes. Luckily for them, it hadn't been the dishes they were washing, rinsing, and drying; rather, the crash came from somewhere else.

"Oh, no-a!" Giovanni exclaimed. "It-a sounds like one of the waiters dropped a tray out in-a the dining room."

Within seconds, Vittorio came exploding into the kitchen through the double-swinging doors, holding a very scared-looking waiter by the ear.

"What kind-a ninny hammer are you, huh?" Vittorio shouted at the waiter, letting go of his ear. "You must-a forgotten rule-a number seventy-one: Use-a your neurons to keep you from dropping trays full of dishes on paying customers!"

"You forgot-a that neurons are-a the basic unit of-a the nervous system. They consist of a nerve body and-a two types of nerve fiber: the dendrites and-a the axons. The dendrites receive-a the signal and send it to the cell body which then sends it out along the axons. There are three-a main types of-a neurons: sensory neurons, which-a carry sensations from-a the body to-a the brain; there are also association neurons, which can-a pick up and interpret information and these are found in-a the brain and spinal chord; and-a motor neurons, which-a carry instructions from-a the brain to-a the body. Nerves are bundles of nerve cells that work together. They contain all of-a the same type of-a neurons or can be a mixture of-a sensory and motor neurons. If they were laid end to end, the body's nerves would-a wrap around the earth twice."

Vittorio stopped and looked around at the silent group in the kitchen. The poor waiter, even though he was a grown man, burst into tears and ran out the back door.

"Oh, great!" Vittorio said, holding his hands up in vexation. "We were already short on-a waiters! What are we going to do now-a?"

The big chef looked around the kitchen at his staff. His eyes roamed until he spotted Blaine and Tracey standing in the back.

"You, two-a," he pointed.

The twins' hearts sank in fear. They both held up their hands and pointed at themselves as if to ask the chef, "Who us?" Vittorio Benaneli reached in a cabinet, and pulled out two clear plastic bags that looked to be full of new folded-up clothing. Then he tossed the bags across the kitchen to the Sassafrases. The twins caught the bags and just stared at them, not sure exactly what they were supposed to do.

"Go put-a those on, and then-a follow me out to the dining

room-a."

Vittorio then did a quick one-eighty and again exited the kitchen.

"Wow! Holy ravioli!" Giovanni exclaimed with his ever-present smile. "There are wait staff uniforms in those-a bags. The two of you are now-a waiters, my friends!"

Shivers Down My Spine

A few minutes later, Blaine and Tracey both emerged from the locker room area, looking like real waiters. They were both wearing tuxedo-style shirts, complete with studs and bowties. They also had crisp black aprons tied around their waists.

"You guys can do it," Giovanni reassured his new friends as they walked past him on their way out to the dining room. "It's-a all about passion!"

Neither twin felt very passionate or excited as they slowly pushed through the double swinging doors. The moment they came through, Chef Benaneli was right there to meet them.

"Being a waiter is simple," Vittorio shared with the twins. "You just-a take-a the trays I give you to the tables that-a I direct you to, okay?"

The twins nodded, understanding as they gulped.

"You just got-a remember rule-a number seventy one-a: Use-a your neurons to keep you from dropping trays full of dishes on paying customers." Vittorio reached up and pulled a huge oval-shaped tray out of a nearby serving window. He handed the tray, which was completely covered in dishes, to Blaine, who somehow managed to balance it on one shoulder and one palm.

"Take-a this to table number fifteen," the chef instructed the boy, pointing to a table in the far corner of the dining room.

He then handed a huge tray brimming with food to Tracey

before saying, "And you take-a this to table number nine-a." Vittorio pointed to a table right in the middle.

The twins both took a deep breath before carefully heading out with their heaping trays full of Italian food. The dining room was decorated very elegantly. Every table was covered with a starched white cloth and had a gently flickering candle placed in the middle. All the patrons were dressed nicely; most were sitting and talking quietly. It was the kind of restaurant the Sassafras twins had seen in movies but had never actually been to. Now, here they were, serving food in a fancy place like this. They just hoped they could follow the rules, especially rule number seventy-one.

Tracey reached her table first. She carefully put her tray down on a tray stand and served the party of four their respective dishes. She didn't spill a drop; she didn't break a dish, so she was relieved. Blaine also managed to deliver his tray full of food to the corner table without any mishaps. They smiled kindly to the patrons and then returned to the serving window. Chef Benaneli already had more trays ready for them to take out.

"You two-a did a great job! Now-a do it again!"

The twins loaded the trays on their shoulders and headed back out into the dining room. They did this over and over again, until they were actually enjoying it more than worrying about it. Everything was going splendidly until the front door to Benaneli's restaurant suddenly swung open and the two meanest-looking Italians anybody had ever seen strolled in.

One was short and thick like a tree stump, and the other was tall and skinny like a flag pole. Both had dark rings around their eyes, menacing grins, and wore chef's uniforms. The atmosphere in Benaneli's completely changed the moment they entered. Food stopped being eaten, conversation stilled, and all was quiet as a chill rested in the air. The tall skinny man pulled out some kind of whistle and started blowing it to get everyone's attention, even though they already had it. The whistle let off a terrible screeching, high-pitched sound that felt like it was damaging their ears. He blew the whistle for at least thirty seconds and then finally stopped.

The short, stumpy man then spoke. "Well, good-a evening, everyone! Is-a everybody enjoying their dinners-a?"

All patrons remained silent and speechless, but Vittorio came barreling out from the back and walked right up to the two in a huff. "What is the big idea? Don't-a you two know about-a rule number ninety-five? No blowing whistles in-a my restaurant!"

The short tree-stump man didn't seem at all intimidated by Vittorio's size or aggression. "Oh, Vittorio, Vittorio, Vittorio," the man responded, "enough of-a your rules-a."

He then pushed Vittorio aside and addressed Benaneli's customers.

"Ladies and gentlemen-a," the man started, with his arms outstretched toward the small crowd. "Have you not had enough of-a the food served here? The same old-a sad dishes, the boring atmosphere-a, and this big-a fat executive chef with-a all of his-a rules?"

Everyone remained quiet, except for Chef Benaneli. He was making a boiling noise, as his face got more and more red by the second.

"You all should get-a out of the Benaneli's rut," the short man continued. "Come to my restaurant-a, which opens tomorrow night, right next door-a!"

This news was too much for Vittorio, who took his chef's hat off and threw it on the ground. He then bent down and, nose-to-nose, started arguing with the short man in sentences so rapid the twins couldn't understand them. As the two were arguing, the Sassafrases noticed that the tall skinny man was pulling something out of his apron. It was a black box, which he set down on the ground and opened up. As soon as the lid was off, the twins saw a mass of small bugs scurrying out across the dining room floor.

Blaine and Tracey looked at each other with wide eyes. What had that man just done? Vittorio hadn't noticed because he was too busy arguing with the shorter man. None of the customers seemed to notice, either, because they were watching the two argue. But that wasn't all the tall skinny man had in his apron. Now, the twins saw him pulling out a small vial of some kind. He snapped the vial in half and dropped it on the floor. Almost immediately, the restaurant was filled with a disgusting pungent odor. As soon as the odor filled the room, the short man pulled away from his argument with Vittorio and addressed the people again.

"Oh, my goodness-a!" he shouted. "What is that awful smell-a? What kind-a restaurant are you running here, Vittorio?"

Then, in sudden succession, table by table, people started screaming.

"There is a roach in my food!" the twins heard a lady shout.

Chaos broke out in the dining room, as all the patrons scrambled up from their tables and started running toward the door to escape from the bugs and the bad smell. Chef Benaneli just helplessly watched as all his would-be paying customers ran screaming out the front door of his restaurant. When everyone was out, the short man shook his head.

"What a shame-a, Vittorio," he said sarcastically. "You should take better care of your restaurant, uh? I don't think it's gone-a be too hard to put you out of business-a."

At that, the two sadistic chefs turned and left, leaving Vittorio and the Sassafrases standing alone in a stinky, roach-infested dining room.

"Who were those guys?" Blaine asked the executive chef. "They were creepy."

"Yeah," Tracey added. "They sent chills down my spine-a."

"That was-a Salvatore and his sous chef, Bruno," Vittorio answered, sadly. "They have been-a trying to put me out of business for years-a. They claim that I got a secret ingredient that I won't share-a. They act very confident about opening their own restaurant, but really, they are afraid-a to open it without my supposed secret ingredient-a. They are all bark and no bite-a." Vittorio shook his head disappointedly as he looked around the disheveled dining room.

He walked over and propped the door open to air out the place a bit. Maybe Salvatore and Bruno were all bark and no bite, but they sure had scared the twins.

"Those two-a do send chills down your spine, don't they-a," Vittorio said, much more slowly than normal. "You know, the spine-a, or actually the spinal cord, is an extension of the brain-a. It runs from the brain to the lower back-a. The nerves that are a part of it relay information from the brain to the body-a."

As Vittorio started talking about the spinal cord, he started talking faster again, and some of his feistiness returned. Just like all the other local experts Uncle Cecil had connected them with, Vittorio Benaneli sure loved his science.

"Thirty-one pairs of spinal nerve bundles branch off of the spinal cord and go to almost every part of the body-a," Vittorio continued. "These spinal nerves are arranged in pairs-a: one for the left side of the body and one for the right side-a. The spinal cord is also responsible for reflexes-a. Like-a if a chef touches a hot pan back in-a the kitchen, the nerve impulse goes from the fingertip to the spinal cord via a motor neuron-a. It tells the chef to pull his hand away immediately-a. His arm-a will jerk away from the hot pan, and the message will reach the chef's

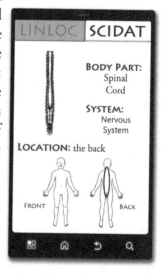

brain-a, meaning he will feel-a the pain. You should-a also know that-a the spinal cord is protected by the spinal column-a. It is no thicker than a finger-a."

The big chef paused and looked around the dining room. "Now, enough about-a the science. We got-a get this place cleaned up and-a cleaned out. I'm not-a gone-a let the mistakes that-a happened here tonight shut me down!"

The twins wanted to tell Vittorio that it was Salvatore and Bruno who had sabotaged the dining room—not any mistake of his or any of his staff. But the big chef left the room in a flurry of purpose before they could relay this information. Blaine and Tracey each smashed a couple of cockroaches as they pulled out their phones to enter their data. They chose a picture for the brain and spinal cord before typing in the information. Then they put their phones back into their pockets and went in the back to see if they could help.

How embarrassing had that display of incompetence been, he thought to himself as he sat down in his basement at 1108 N. Pecan Street, bemoaning his last failed attempt at thwarting the learning adventures of those determined twins. He had "borrowed" the Dark Cape suit, hoping to use its disappearing capability to somehow bring the Sassafras twins to ruin, but he never did manage to figure out how exactly it worked. It would make him disappear when he wanted to be visible, and when he wanted to be visible, it would make him disappear. He never gained any control over it.

Another thing that had been disappointing about this last outing is that those twins didn't seem to be scared of him. Instead of running from him, they chased him and nearly caught him. Even worse than the Dark Cape debacle had been getting arrested in Ethiopia. Luckily, he had been able to use his carabiner and invisible zip lines to escape custody. He really needed to come up with some better ideas.

He knew that right now the twins were in Venice, Italy, gathering information on the nervous system, but he hadn't yet calibrated his three-ringed carabiner to go there after them. He was fresh out of ideas on how to stop them. Yes, the outlook for revenge on Cecil Sassafras was bleak at the moment, but that didn't mean he was going to give up—not at all.

CHAPTER 7: EXPLORING THE SENSES

Touch and Sight

The smell was gone, the roaches were gone, and the restaurant was clean from front to back. It had been a long night of cleaning, and the Sassafras twins had been so tired they had just fallen asleep right there in the dining room, on the cushioned seats at a booth table. Tracey was having a fascinating dream about the different parts of the nervous, respiratory, and skeletal systems all turning into King penguins, and then working together to do an amazing choreographed dance routine. Blaine was dreaming about speaking an actual coherent sentence around Julie Ette.

The two were snoozing blissfully, unaware that someone had just entered the dining room and was walking toward their booth right now. Blaine was shaken awake first, then Tracey. It was Giovanni, and he was very excited, even though it was still very early in the morning.

"Good morning, you two!" he exclaimed, with a big smile. "You got-a get up! We got-a great chance-a today. Uncle Vittorio is-a going to the countryside-a, and he has invited us to come along with him-a!"

Blaine and Tracey weren't sure why going to the countryside was such a great chance, but it sure was making Giovanni happy so they figured they had better get up and join him. They each sat up, rubbed their groggy eyes, and followed Giovanni as he bounded happily out of the dining room. The three joined executive chef Benaneli, who seemed to still be discouraged this morning, and headed out to the canal. They walked down a few quaint little alleys and over a couple arched bridges until they arrived at a place where several gondolas were docked. Chef Benaneli paid the fare, the gondolier stuck his long paddle down in the water, and off they floated gently through the picturesque canals of Venice.

"Well-a, kiddos," the big chef stated sadly, "I think maybe I have-a lost touch with-a reality. Not-a literal touch, which is one of-a the five senses. Your skin is-a responsible for giving your brain a 'touch picture' of-a the world around it. The skin has millions of-a sensors that-a detect pressure, vibration, temperature, and-a pain. Some parts of the body are more sensitive to-a touch than-a others. The hands and-a lips have the highest concentration of-a touch receptors, while-a the arms and-a legs have-a the least. Once-a the sensor detects touch, they send-a their message via nerve fibers to the brain-a."

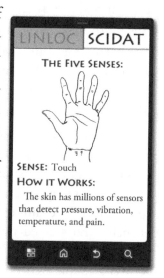

LINLOC **SCIDAT**

THE FIVE SENSES:

SENSE: Touch
HOW IT WORKS:
The skin has millions of sensors that detect pressure, vibration, temperature, and pain.

"No, that is not-a the kind-a touch I am talking about. I thought I had-a the best restaurant in all of Venice. But I must have been blind in-a my pride. All I have is-a stinky, bug infested spaghetti house! I forgot-a rule-a number fifty-six-a: Don't lose-a touch with-a reality!"

Blaine and Tracey couldn't believe that Chef Benaneli still thought the roaches and the stink bomb were his fault. They had to tell him the truth.

"Chef Ben—" Blaine started to say, but he was interrupted by the passionate Giovanni.

"Uncle Vittorio!" the fourteen-year-old boy said, standing up from his seat. "You have not-a lost touch with reality. You are-a the smartest-a man I know! Just-a listen to all of-a the science you are teaching us right now! Plus, you are the best-a chef I know, too! Benaneli's is-a the best restaurant in all of Venice! Maybe it is-a your staff's fault. Maybe we have-a failed you! Maybe we have-a lost sight of your vision!"

"No, boy, no! Sit-a down, sit-a down!" Vittorio pulled his sincere nephew back down to his seat. "I know I am-a hard on all of you, but you-a are the best-a staff a chef could-a ask for. You have-a not lost sight. That's not-a what sight is, anyways. Sight is-a the body's dominant sense. Your eyes are what allow you to-a see. Light enters the eye through-a the pupil. The pupil changes size based on-a the amount of-a light. It-a gets bigger for less light, and-a gets smaller for more light. After the pupil, light passes through the cornea and lens of-a the eye, which focuses the light and-a produces an upside down-a image of-a what you are seeing at-a the

UNLOC SCIDAT

THE FIVE SENSES:

SENSE: Sight
HOW IT WORKS:
Sight is the body's dominant sense. Your eyes are what allow you to see.

back of your eye on-a the retina. Then-a the sensor cells in the retina called photoreceptors that-a are stimulated by light change the image into an electric message which is sent to your brain along-a the optic nerve. The brain processes the image so that-a you can see it right-a-side-up. There are two types of photoreceptors in-a the retina: rods, which-a work best in-a dim light because they are sensitive to black and-a white, and cones, which work best in-a the bright light because they detect color.

"That-a, my dear nephew, is-a what sight is and the staff at-a Benaneli's has not-a lost it. Last-a night's debacle was-a all my fault. That is-a why we are taking this trip to-a the countryside. I need to get away and-a clear my head and-a refocus. And Giovanni, you know who lives in-a the countryside!"

The Italian boy's eyes got wide with excitement at Vittorio's last statement. "Mama Mia!" he exclaimed. "Are we going to-a Grandmama's house?"

"That's-a right, Giovanni!" Chef Benaneli boomed, with a smile hidden under his mustache. "We are going to-a Grandmama Benaneli's house-a! My mama, your grandmamma, and-a the best cook in-a the whole world!"

Giovanni looked at the twins with his biggest smile yet. "Blaine-a! Tracey! You are gonna love it! Grandmama Benaneli is-a the best!"

The twins smiled as the gondola continued its way through the maze of beautiful waterways. It was fun to see Giovanni so happy and excited, and also to see a softer side of Chef Benaneli. Their gondola was almost out of the smaller canal and out onto the grand canal when a look of concern came over Vittorio's face.

"Uncle Vittorio," Giovanni inquired, "what's-a wrong? You look alarmed."

The big chef pointed behind them, "It-a looks like-a we are being followed!"

The three children turned their heads, and their hearts jumped up into their throats as they saw that it was true. A speedboat was barreling up the small canal straight for them. And who could be seen peeking over the bow, except for Salvatore and Bruno!

"It's-a those two bad guy chefs-a!" Giovanni exclaimed. "Why are they chasing us?"

"They must-a found out about-a my plans to go to-a the countryside to visit Mia Momma," Chef Benaneli yelled. "They think I cook-a with a secret ingredient, and they want-a to steal it. They have-a no creativity of-a their own!"

The two bad chefs' faces looked as mean as ever as they plunged their speed boat full-speed ahead, straight toward the four's gondola. Though they were almost to the wider waters of the Grand Canal, the Sassafrases weren't sure they were going to make it out of the smaller canal before Salvatore and Bruno's big speedboat smashed their smaller boat into pieces. The gondolier, sensing the impending danger, stroked harder and faster with the paddle. Immediately, they started going a little quicker, but nowhere near as fast as the approaching speedboat. Would they make it?

The roar of the coming engine got louder and louder. The gondolier stroked faster and faster, but the bow of the speedboat got closer and closer. The nose of the gondola had now made it out into the Grand Canal. Now the gondolier just needed to cut the gondola sharply around the stone wall corner to avoid the chasing boat. He skillfully and powerfully cranked on the paddle, pushing the gondola sharply to the left.

They had almost made it safely around the corner when Salvatore and Bruno's boat clipped the gondola's stern. Though not much contact was made between the two boats, the force of the bigger speedboat hitting the smaller boat sent the back of the gondola jerking upwards. The twins felt themselves floating off of their seats and up into the air. They flailed their arms and legs helplessly as they soared through the sky over the water. It all seemed to happen in slow motion, as they saw Vittorio and Giovanni floating through the air with them. They reached their highest point and then started to fall back down towards the water, but before their bodies hit the surface of the river a separate speedboat came zipping by and caught them like a basket. Vittorio and Blaine landed on a seat, while Tracey and Giovanni both bounced off the railing of the boat. They went sprawling over the side. Blaine instinctively reached out and grabbed a hand. It was Giovanni's, but what had happened to Tracey?

Blaine grasped Giovanni's hand as tightly as he possibly could. He then pulled himself up to his knees, leaned over the side of the boat, and reached out and grabbed Giovanni's other hand. The Italian boy was clinging to Blaine for all he was worth. Blaine frantically looked out over the water behind them for his sister, but she was nowhere to be seen.

"What happened to Tracey?" Blaine shouted the question to Giovanni over the roar of the boat.

"Look at-a my leg!" Giovanni hollered back.

Blaine looked down, and there was Tracey with both hands firmly around one of Giovanni's ankles. She looked scared and was obviously completely soaked, but it looked like she had a vise-grip hold on Giovanni's legs; Blaine knew she wouldn't let go. Now he just had to figure out how to pull the two back into the moving boat. The driver of this speedboat had a look of utter confusion on his face. He had no idea how he had just caught four human beings who had been flying through the air. He eased off the accelerator and the boat began to slow down. Just as he did, the riders saw a circular splash of water, as Salvatore and Bruno whipped their boat around to continue pursuit. Their bow rose back up in the air as they slammed on the gas and started another beeline straight for the four. Their faces showing that they were very determined to catch Vittorio and his three sidekicks.

The driver of the boat that the four were in had let completely off the throttle by this time. He looked like he was about to ask the obvious questions, but no words were coming to his mouth. Vittorio, who had now recovered from his tumble, reached over the side and with two big strong arms pulled his nephew and Tracey back into the boat.

He then looked at the speechless driver and shouted, "Step-a on it!"

The driver jerked into action, obeying Vittorio's stern command. He pulled back the throttle and off they went, up the Grand Canal with Salvatore and Bruno hot on their trail. There were all kinds of boats meandering through this new canal, and the two speedboats zipped around and past them all. The Grand Canal eventually led out to the Venetian Lagoon, a much bigger and wider body of water. Their driver, more than likely still full of questions, drove full speed ahead. His boat was proving to be a little faster than Salvatore and Bruno's, as they slowly began to inch away from the chasing evil chefs.

As they got close to reaching the other side of the lagoon, Vittorio directed the driver to take them to a nearby dock. When they docked, he paid the driver for being their unlikely wheelman, and the four quickly got out and ran up the dock. They'd put some distance between themselves and the pursuing chefs, but Salvatore and Bruno were still cruising quickly their way. Chef Benaneli moved surprisingly fast for such a big man. Giovanni and the Sassafras twins followed him as he reached the end of the dock and took a right, running up a small street. Tracey glanced back and saw that Salvatore and Bruno had also reached the dock and were already climbing out of their speedboat, with determined faces that were bent on pursuit.

"Man, they really want that secret ingredient, don't they?" Tracey thought to herself. She just hoped they could figure out a way to escape the clutches of the two—and fast. Vittorio led them up through a row of small garages, looking at each as they ran by.

He eventually stopped at one and pulled a set of keys out of his apron. He fumbled through the keys until he found the one that he wanted. He then quickly reached down and stuck it in the lock on the garage door of his choice and turned. The lock clicked open, and Vittorio bent down and yanked the door up, revealing that there, parked inside the small garage, was an old model motorcycle complete with a side car.

Just then, Salvatore and Bruno came into view. They were now running up the row of garages toward the four.

"Quick! Get-a on the bike!" Chef Benaneli commanded.

The three children eagerly obliged, as Blaine, Tracey, and Giovanni all jumped into the sidecar. Vittorio kicked his big leg up and straddled the motorcycle. He then, again, started fumbling through his keys to find the one that started this vehicle.

"Aha," he declared when he finally found it. "There-a it is!"

He reached down to stick the key in the ignition, but just as he did, Salvatore Bruno came running around the corner of the garage door.

"Where-a do you think-a you're going?" the short Salvatore growled.

Bruno remained silent, but he walked up and put his hands on the motorcycle's handlebars. The chefs looked the four over with menacing dark-ringed eyes. Salvatore stepped up and snatched the keys out of Vittorio's hand.

"We plan-a to open our restaurant tonight-a," he growled through gritted teeth. "And we need-a your secret ingredient to do so."

Vittorio didn't say a word, but the twins noticed that his face was growing red.

"Why don't-a you guys get-a your own secret ingredient?!" Giovanni shouted, defending his uncle. "Use-a your own recipes.

Your own brains! Your own-a blood, sweat, and-a tears!"

"Now, why would-a we do that," Salvatore said, still gritting his teeth, "when it is-a so much-a easier just-a to take yours?"

Giovanni threw his hands up in exasperation. "What kind of people are you? Do you not-a know right from-a wrong? Have-a you no mamas?"

A wicked smile formed on Salvatore's face. "I'm-a glad you brought up-a the subject of-a mamas, because we happen to know you are going to Mama Benaneli's right-a now. If-a you won't-a give us the secret ingredient, maybe she will. We have-a ways of making people talk-a. So you are-a going to take us to her right-a now!"

Vittorio's face was now redder than the twins had seen it yet. He'd had enough of these two wanna-be chefs. He stepped off his motorcycle and stood up to his full height. A small look of panic formed on Salvatore's face, as if he was just now remembering exactly how big Vittorio Benaneli was. The big round chef looked down at the short stumpy man.

"Rule-a number three," Vittorio stated, snatching the keys back out of Salvatore's grasp. "Nobody threatens mia mama!"

Chef Benaneli suddenly jumped into action. Moving like a much smaller man, he grabbed a cord that was lying on the garage floor, and he wrapped both Salvatore and Bruno in it—tying them together. He then jumped back on the motorcycle, turned the key, and gave a big kick to start the engine. He pulled the motorcycle out of the garage and into the street. He jumped back off the bike and started to close the garage door, but before he did, he had a sort of promise for Salvatore and Bruno.

"Don't-a worry, you two. We will-a be back to let-a you out. And when-a we do, we will-a have a little surprise for you." Vittorio then slammed the door closed and locked it.

He got back on the motorcycle and, without saying a word to the three in his side car, took off down the street. The Sassafras' eyes were wide with wonder. They could not believe what they had just seen. Chef Benaneli was not a man to be messed with, especially if someone talked about his mama. It was now very apparent that the once intimidating duo of Salvatore and Bruno were not that much of a threat. Blaine and Tracey wondered what kind of surprise Benaneli was going to give them when they returned.

Taste, Smell, and Hearing

Vittorio drove the motorcycle up through the row of garages and then wound around through some city streets until they finally reached the open roads of the countryside. This is where Vittorio kicked the bike up to high gear, and off they went across the rolling hills, enjoying the beautiful morning, sunshine and the wind blowing through their hair. Though they were a tad bit cramped with three people, Blaine and Tracey were loving riding in this side car. Tracey was in the front, Blaine in the middle, and Giovanni in the back, and they were all smiling and waving their hands around in the wind. Vittorio was also smiling, though it was hidden under his windblown mustache.

The Venetian countryside was stunning. They zipped past vineyards, fields of growing vegetables, and ancient Italian homes. Vittorio handled the road and all its smooth curves like he knew them well and had driven them thousands of times. Eventually, he pulled the motorbike off the main road and onto a small dirt road. They followed this road until they reached a small stucco-styled house with a red tile roof. It was surrounded by flowers of purple, yellow, and white, and the front door stood wide open.

Chef Benaneli brought the motorbike to a stop and all four riders got off. The three children followed the big bouncing chef into the home, but before they even entered, Blaine and Tracey noticed that they could smell a mouth-watering aroma wafting out into the air. It immediately made them want to eat whatever it was that was cooking.

As soon as they set foot in the door, Vittorio held out his arms and exclaimed "Mama Mia!"

A small older Italian woman looked up from a table where she was listening to opera and grating cheese. She shouted in an excited and merry voice, "Little Vitty, my baby! And-a Baby Gio, my grandbaby! Have-a you come-a to visit me?!"

"Yes-a Mama!" Vittorio grinned widely. "We have-a come to visit you in-a the countryside-a."

Grandmama Benaneli came around the table and gave 'Little Vitty' and 'Baby Gio' big hugs. Though she had never met the Sassafrases before, she hugged them also, like she had known them their whole lives. She then stepped back and gave the four a once-over.

"You all look like you are-a wasting away. When is-a the last time you ate? Go sit-a down at-a the table. I am just-a making some tortellini. You look-a like you are starving. I must-a feed you!"

She then lovingly shoved and pushed all four over to a dining room table and forced them to sit down.

"Mama, let-a me help you with-a lunch," Vittorio offered.

"Nonsense-a!" Grandmama exclaimed. "You just sit-a there and-a relax. I know you work oh so hard at-a your restaurant-a every day. You need a break. Just let-a Mama feed you!"

The group could tell there would be no arguing with Grandmama Benaneli on this day, not even for her own son, so they all just sat and watched the woman do what she had been doing for years.

"Little Vitty," Grandmama Benaneli addressed her son as she worked, "Tell-a Mama, how is-a everything going at-a the restaurant-a?"

"Oh, Mama," Vittorio said, almost like a big kid. "Things are not-a going so good. Last-a night, we had-a some buggies and a stinky smell. We've-a never had-a those kinda problems before. Maybe I forgot rule-a number ninety-five-a: Keep-a the restaurant clean-a."

The twins still hadn't been able to tell Chef Benaneli that the roaches and the smell had not been his fault. Blaine started to bring it up right there and then, but the big chef kept yammering on.

"And on top of-a that, Salvatore and-a Bruno are opening up a restaurant right-a next door to Benaneli's. They want-a to put me out of business. They claim I got-a some kind of-a secret ingredient and they want-a to steal it!"

"Secret ingredient?!" Grandmama Benaneli roared like even the mention of such a thing was preposterous. "We Benanelis don't-a use no secret ingredients!"

"That's-a what I told them, Mama. But-a Salvatore and Bruno, they don't-a believe me!"

"Salvatore and Bruno?" she asked as she minced some garlic. "Aren't those-a the two that used to make-a fun of you in-a grade school for liking science so much-a?"

"Yes-a, Mama. That is-a them," Vittorio answered.

"Blah!" Grandmama Benaneli exclaimed in disgust. "Those-a two buffoons don't now what-a they're talking about. The Benaneli family has-a been creating fine dining experiences for generations without-a no secret ingredients! We focus on-a five things to make-a every meal a victorious celebration! Littly Vitty, you know what-a those five things are."

"The five-a senses," Vittorio replied.

"That's right," Grandmama Benaneli confirmed. "The five-a senses! Taste, smell, sight, hearing, and-a touch. You focus on-a these, and-a every meal will-a be a success!"

The energetic little Italian grandma looked at Blaine and Tracey. "Has-a Little Vitty told-a you two about-a the science of-a these five senses?"

"Yes, ma'am," Tracey answered. "At least he started to. He told us about touch and sight."

"That's-a my boy," Grandmama smiled. "He's-a so good with-a the science. Vitty, tell-a these kids about-a the other three-a."

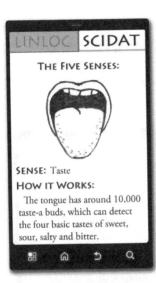

SENSE: Taste

HOW IT WORKS:
The tongue has around 10,000 taste-a buds, which can detect the four basic tastes of sweet, sour, salty and bitter.

"Yes-a ma'am," Chef Benaneli said. "Let-a me start with-a taste. The tongue has around-a ten thousand taste-a buds, which can-a detect four basic tastes: sweet, sour, salty, and-a bitter. These-a taste buds are located at-a the tops of-a the papillae—the bumps that cover the tongue. The taste-a buds detect chemicals in-a the food. When-a the chemical enters the pore of-a the taste bud, they stimulate the bundle of-a sensors inside, and-a the sensors then send a message to-a the brain. This allows the person to-a taste the food that they are eating. The tongue has-a four different regions of-a taste buds. The front of the tongue tastes sweet. Behind-a that is salty, and behind-a that is sour. And at the very back are-a the taste buds for-a bitter tastes."

The Sassafras twins couldn't wait to use their sense of taste on the tortellini that Grandmama Benaneli was finishing up even now.

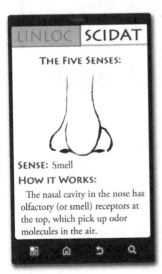

SENSE: Smell

HOW IT WORKS:
The nasal cavity in the nose has olfactory (or smell) receptors at the top, which pick up odor molecules in the air.

"The sense of-a smell and taste work-a closely together." Vittorio continued, "Both-a detect chemicals. The nasal cavity in-a the nose has-a smell receptors at-a the top, which-a pick up odor molecules in-a the air. The receptors then-a send the message to-a the brain. There are about ten million olfactory receptors in-a the roof of-a the nasal cavity, which allows humans to-a detect almost ten thousand different odors. Smell dominates over taste when it comes to food, which is-a sometimes why cold-a food can-a be tasteless."

Blaine and Tracey didn't know how many of the ten thousand odors that their olfactory receptors were picking up right now here in Grandmama Benaneli's house, but however many there were, they were all good. Their mouths began to salivate in anticipation of the upcoming feast.

"And-a lastly, let-a me tell you about hearing," Vittorio shared. "The ears allow-a the body to hear a range of-a sounds. Sounds travel in a wave through the air until it-a reaches the ear. The port of-a the ear that-a we can see, called the pinna, funnels the sound into a canal. At-a the end of-a the canal there is a taut piece of-a skin called the eardrum. The eardrum vibrates as-a the sound wave hits it. These vibrations then-a pass through-a the three ossicles of-a the middle ear. These-a three ossicles are-a called the hammer, anvil, and-a stirrup. They transfer the sounds to-a the fluid filled inner ear. The movement of-a the fluid is then detected by-a sensors in-a the inner ear, which-a change the movement into electrical impulses that are-a then sent to-a the brain. The louder the sound, the larger the vibrations and-a movement in-a the inner ear."

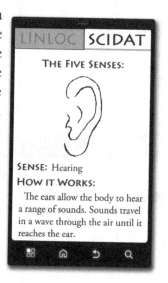

Just as Vittorio finished giving his information, Grandmama Benaneli slid hot plates of scrumptious looking tortellini onto the table in front of each of them. Giovanni grabbed his fork and started to dive right into his ring-shaped, meat-and-cheese-filled pasta, but he was interrupted by his grandmother.

"Baby Gio, before we eat, we say grace-a," she admonished with a smile on her face. Grandmama Benaneli bowed her head and prayed, ending with, "Amen-a."

Giovanni again tried to quickly stick a fork into a tortellini and shove it in his mouth, but this time he was stopped by his uncle who grabbed his wrist.

"Wait a second, mia nephew," the big chef chided. "Just-a take pause and-a enjoy this-a moment. We are about-a to eat a perfect meal. Mia Mama has-a led by example as-a always. Take note of-a how the five senses are being indulged right-a now. Touch: the nice-a heavy forks in-a our hands and the perfect texture of-a the pasta. Hearing: Pavarotti can-a be heard singing softly opera on-a the radio in-a the background. Sight: the tortellini has-a been folded just-a perfectly for an attractive look, and has-a been covered with the perfect amount of sauce to make it-a look very pleasing to-a the eye. Smell: the herbs and-a spices used in-a the dish have-a been chosen because they are inviting and memorable. Finally, taste: well that is-a enough talk. Let's eat!"

Vittorio let go of his nephew's wrist, and within seconds Giovanni had a tortellini in his mouth. Blaine and Tracey quickly followed suit, and as soon as those rings of pasta hit their mouths, they were overcome with complete satisfaction. It was quite possibly the best food they had ever eaten.

The Sassafrases enjoyed the company of the Benaneli family and Grandmama Benaneli seemed to really enjoy having people out in her countryside home. Vittorio was recharged and refocused again. Giovanni was as passionate as ever. It had been a good visit. Grandmama Benaneli had an almost magical ability to encourage with words and with food.

"Mama, I have a surprise for you," Vittorio announced as the four got up to leave.

Minutes later, Grandmama Benaneli was riding behind her son on the motorcycle, and the kids were again cruising together in the sidecar. It was now early afternoon, and the sun was still shining brightly, making it another enjoyable ride through the rolling hills of the Venetian countryside. Eventually, they reached the row of garages down by the lagoon. Vittorio brought the motorcycle to a stop in front of the appropriate door. He then hopped off the bike, unlocked the door, pulled it open, and there were Salvatore and Bruno, still tied up with the cord. They didn't look quite as mean and menacing now as they had last night and this morning. Evidently, their time tied up in a dark garage had humbled them a bit.

"So, Mama," Vittorio declared, "these are-a the two men I was-a talking about. They wrecked our gondola, threatened you, Mama, and they want-a to steal our secret ingredient."

Grandmama Benaneli stepped into the garage, walked right up to the two bound wanna-be chefs, and pulled them up to their feet by their ears. "You-a two have no business opening up a restaurant-a if all you want-a to do is steal other people's ingredients. Being a chef is about-a being creative, not-a being a copycat. It's-a also about paying close attention to-a the basics. It-a sounds like-a you two need-a some discipline."

Salvatore and Bruno looked like they were about to cry. Though Grandmama Benaneli was small, she was intimidating. There was something about the tone in her voice. It let people know that there was no opposing her.

"I'm-a going to untie you, and-a then you are going to-a come with me. If you two really want-a to be chefs, you have a lot of work-a to do," Grandmama remarked to Salvatore and Bruno. "We will start with chopping onions and-a peeling potatoes. Do you understand-a mama?"

"Yes-a ma'am," Salvatore and Bruno both agreed with scared sniffling voices.

THE SASSAFRAS SCIENCE ADVENTURES

Grandmama Benaneli loosened the two from their cords, and then she led them by their ears again—this time to the motorcycle. She put them in the sidecar, then she jumped on the seat and kicked the machine back to life. She gave a big happy wave to her son, her grandson, and the Sassafrases. Then she took off, riding the motorcycle at full speed, back out into the countryside.

Vittorio looked at the three children. "Well, I guess-a we don't have-a to worry about-a Salvatore and Bruno anymore."

CHAPTER 8: COACHING IN BEIJING

Boiling Blood

The Sassafras twins sat at a quiet table in Vittorio's restaurant, entering the remaining information into their SCIDAT app. They also flipped through the archive app to find the appropriate pictures to send to Uncle Cecil along with the information. They were in good spirits, but both were very tired after another night of serving as waitstaff at Benaneli's. After getting rid of Salvatore and Bruno, they had hired a boat to return to the restaurant with Vittorio and Giovanni. Chef Benaneli had been almost giddy after his rejuvenating visit with his mother in her countryside home. He was refocused and as determined as ever to give the patrons at his restaurant the best dining experience of their lives. By getting back to the science of it, and simply focusing on the five senses, he was able to deliver the perfect meal that night to his delighted guests.

Blaine and Tracey never did get the chance to tell Chef Benaneli that it was Salvatore and Bruno that had produced the bugs and smell in the restaurant the night before, but at this point they figured it didn't really matter. Everything had actually worked out for the good of Vittorio. It had worked out well for the Sassafrases too. They had learned more about the human body, made new friends, and lived through another exciting leg of their scientific travels.

"OK," Blaine sighed, groggily. "I think I found a good picture from the archives to represent hearing." He held his phone up to show his sister.

"Yep, that's a good one," Tracey answered. "And I just finished sending in all the data. So are we ready for LINLOC?"

"Looks like it," Blaine responded, as he hit send on his

SCIDAT app.

The twins switched over to the LINLOC app on their smartphones to see where they would be going next, who they would be meeting, and what they would be learning about. As soon as the information popped up on the screen, Tracey smiled.

"Well, dear brother, it looks like now it's my turn to host you in a country that I have been to, but you haven't," she declared.

"China!" Blaine exclaimed.

"That's right," Tracey affirmed. "But, this time, the destination is not Sichuan. It's Beijing: Longitude 116° 40' E, Latitude 39° 91' N. Our local expert will not be Tashi Yidro, but someone named Coach Boxton."

"Our topics for study will be the circulatory system and the muscular system," Blaine added, reading the screen.

"That sounds fun," Tracey yawned, "but I think it can wait until tomorrow."

"I agree," Blaine nodded as he also yawned.

At that, both twins lay down right where they were, at the booth table, and fell asleep almost immediately.

Early the next morning, before anyone had come into the restaurant, the twins opened the LINLOC app on their phones once again. It was time to zip! The light around them seemed almost tangible, like you could reach out, grab it, and hold it in your hands. It wasn't like light that you simply looked at, but it was light that you were encased in. The twins were pretty sure they would never lose their awe for this speed of light zip line travel thing. They landed with a jerk, having just left Venice, Italy, and now presumably had arrived somewhere in Beijing, China.

They had landed, but they were still moving. This had happened before, when they had landed on the back of one of Princess Talibah's camels in Egypt, but this was no animal they had landed on. It was some kind of metal vehicle, and it was moving fast. The twins nervously anticipated their sight and strength returning. They were hoping these faculties would help them hang on to whatever it was they were on top of. They didn't know how high or low they were off the ground, but they didn't want to find out by falling off.

Tracey felt her body starting to slide. Her leg fell over the side and dangled in the air, threatening to bring the rest of her body down with it. She tried to reach out and grab something, anything, with her hands to hold onto, but she just didn't have enough strength yet. Her sight slowly returned, and she saw that she and Blaine were indeed on top of a piece of moving metal, but she couldn't tell exactly what it was. Now her other leg fell over the side. She was lying on her stomach, now just her torso remained on top of the vehicle, and it was sliding off too.

"Aaarrrgh!" Tracey screeched in determination, as she reached out and grabbed Blaine's shirt, virtually willing her strength to return before she fell over the side. It worked, and she clung to her brother's cotton tee for all that she was worth.

As soon as Blaine's strength returned, he reached down and grabbed his sister's arm with his right hand to pull her back up, and with his left hand, he held tightly to the front edge of this speeding metal roof they were on. It wasn't a big roof, really—just big enough for the two of them to fit on top of it. It wasn't a car or a truck, but it was moving through the busy street of what was, most definitely, Beijing. And it was going fast!

"I think we've landed on top of a rickshaw," Tracey cried.

THE SASSAFRAS SCIENCE ADVENTURES

"I thought rickshaws were pulled by people on foot, or maybe pedaled, like a bike," retorted Blaine. "This thing is going way too fast to be a rickshaw."

"Well, obviously it has a motor attached," Tracey explained. "But I think it's a rickshaw."

The twins hung on side by side as the vehicle they were on top of weaved in and out of traffic. Traffic that was comprised of cars, motor scooters, bicycles, and people—lots and lots of people. The Sassafrases couldn't figure out how whoever it was that was driving this thing was avoiding a crash. As they hung on, they also began hearing a conversation between two people that were riding in the speeding vehicle. A conversation that sounded somewhat heated.

"I said, 'No'," a deep strong voice bellowed.

"But you said you need help with your decathletes," a whiny, sing-songy voice said, responding to the 'no.'

"I do!" the first voice confirmed. "They didn't perform

in the London games even close to their capability. I think my decathletes were just lacking a little determination and diligence, but I absolutely will not do what you are suggesting, Dr. Veeginburger."

"Why not, coach? It's easy, it's cheap, and it will make your athletes perform at twice their capability. Simply inject this medicine into their bloodstreams once a week, and then sit back and behold the results. It will be mind-blowing."

"No, no, no!" The coach responded resolutely to the doctor's sales pitch. "The decathletes of China are not cheaters! And their blood will not have your despicable chemicals floating in it!"

"C'mon, Coach Boxton," the whiny doctor moaned, still trying to sell his supplements. "My potion can help you. What do you really know about blood, anyway?"

"What do I know about blood?" asked Coach Boxton, as if he was accepting the challenge. "The blood is a liquid that carries substances to and from the cells in the body. It also distributes heat throughout the body."

"Coach, I didn't actually want to hear you explain what blood is," Dr. Veeginburger pleaded, trying to stop the coach, but Boxton continued on nonetheless.

"Blood is composed of four parts: the plasma, the platelets, the red blood cells, and the white blood cells. Plasma delivers the nutrients the body needs to the cells. It is the liquid portion of the blood. Platelets help to heal wounds. They stick together and form a temporary plug in the gap that a cut causes in a blood vessel. And blood vessels, by the way, are the tubes that carry blood around the body. Red blood cells deliver the oxygen the body needs to the cells. They are shaped like doughnuts. And they are the most numerous cells in the body, numbering over two hundred fifty million in just one drop of blood. White blood cells play a part in the body's defense against infection. They track down bacteria and destroy the germs."

Suddenly, the vehicle took a sharp turn to the right, nearly throwing both twins clean off, but both Blaine and Tracey somehow managed to reach out and regain handholds on the roof of the speeding carriage.

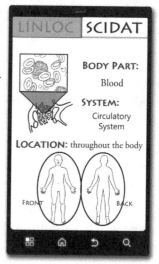

"The common adult has an average of nine pints of blood, which circulates through the body continuously," Coach Boxton said, down in the vehicle that the Sassafrases were barely hanging onto.

"Finally, Dr. Veeginburger, when blood has a lot of oxygen, it is bright red, but once it has unloaded its oxygen, it turns a deeper, darker shade of red. If you let it sit, it will separate into yellow liquid plasma and also into layers of platelets, white blood cells, and red blood cells. In other words, Doc, the blood is like an important part of an intricately designed machine that works fine without any of your chemicals floating around in it. What we need is just a little more determination and diligence. However, what you need, Dr. Veeginburger, is to get out of my rickshaw."

"So it WAS a rickshaw, after all," the twins thought to themselves. Evidently, in China, rickshaws can be motorized.

The driver of the rickshaw abruptly put on the brakes, and the vehicle screeched to a stop. The sudden inertia threatened to throw the twins from their perch, but again, they managed to hang on. Dr. Veeginburger, however, was not so lucky. The twins could now see him as he was thrown out of the rickshaw by a pair of big, strong-looking hands. The doctor was a tiny little man who, in a strange way, looked a little bit like a mouse.

"I never want to see you again, doctor," Coach Boxton hissed from inside the rickshaw. "Stay away from me, and stay away from my athletes. We do not want, and will never want, any of your illegal supplements! Be gone with you!"

With that, the rickshaw started back up and took off, leaving the shady doctor standing by himself on a curb somewhere in Beijing. After a few more minutes of swerving and weaving through the busy, ever-moving traffic, the rickshaw stopped again—this time in front of several impressive looking buildings. One looked like the huge legs of a giant. One looked like it was all wrapped up in lights and bubble tape, and yet another looked like a big bowl made out of crisscrossed, interwoven steel beams. Coach Boxton was a big strong man about the same size and stature as Nicholas Mzuri, the twins' first local expert. He got out of the rickshaw and paid the driver. He was about to turn and walk away when he spotted the twins. His left eyebrow dropped, and his right eyebrow rose up on his forehead.

"What on earth are the two of you doing on top of my rickshaw?" the big coach asked, in a deep, perplexed voice.

Surely this would work. This was his most successful project since the robot hummingbird. He had left the kids alone while they were in Venice so he could perfect his latest invention. Okay, maybe it wasn't completely his invention because it was loosely based on a model that Cecil had designed years ago, but he had tweaked it. It was now what he called an Expandable Trap Box. Nothing he had done so far had worked in stopping those Sassafras twins, so he knew it was time to get more creative. He was sure this box would be the thing that finally brought an end to the twins' science learning.

The way it worked was simple. The box started small—small enough to fit into your hand. But when you threw it, it rapidly expanded. It not only expanded, it also opened up and trapped whatever you were throwing it at. Then it would lock itself. It was sound proof, escape proof, and fool proof. This is how he would catch and stop those twins.

He had completed two boxes, one for each twin. He knew they were now in Beijing, and in just a few minutes he would be there himself. He strapped on his harness, calibrated his carabiner, and grinned a confident, wicked grin as he slipped both palm-sized expandable trap boxes into his backpack.

Hearts of Gold

"Hello! My name is Jek. Welcome to the Beijing Olympic Village."

"Jack?" questioned Tracey.

"No, I think he said Jake," Blaine corrected his sister.

"Sorry," Tracey smiled. "I didn't quite understand you. Is your name Jack or Jake?"

"Yes!" the teenaged Chinese boy stated, happily. "That's right, and I am Coach Boxton's translator."

"That he is," Coach Boxton agreed. "He is the best translator in all of Beijing!"

The twins looked at each other, confused but happy. They were just glad they were standing here at all. When the big coach had spotted them on top of his rickshaw, both were pretty sure that Boxton was going to toss them to the side and leave them in the dust, just like he had the shady Dr. Veeginburger, but instead, he had invited them to come check out the Olympic Village with him.

"The Beijing Olympic Village is also known as 'The Olympic Green'," Jek informed as he pointed to the building that looked like a bowl. "Now let's go to the Bird's Nest, which is the most famous building in the Olympic Green. This is where we will find Coach Boxton's decathletes."

The twins followed the big Coach Boxton and the little Jek across the Olympic Green to the Bird's Nest. Once there, they took an elevator up and entered the building through an upper level entrance. When they stepped out, the Sassafrases were amazed at what they saw. The Bird's Nest was a huge sport's arena that was beautifully designed to hold every event in track and field. About a dozen or so athletes could be seen down on the field, engaging in several different track and field disciplines.

"These are my decathletes," Coach Boxton proudly declared. "Or, more precisely put, they are China's fine Olympic hopefuls for the decathlon in the next summer games. I am honored to be their coach."

The twins could tell by the smile on his face that Coach Boxton was very proud of his athletes, but one thing puzzled them. How could Boxton be the coach of the Chinese decathletes when he wasn't even Chinese?

As if reading their minds, Jek interjected, "Even if you aren't Chinese, we are honored to have you, too, Coach Boxton!"

Jek turned to the twins and expounded, "Here, in China, we care very much about the Olympic games. We want to get better and better in every sport. So we find and hire the best coaches in the world to instruct our athletes. Coach Boxton just happens to be the most capable decathlon coach on the planet!"

The big coach chuckled as his translator flattered him, and he shook his head. "Determination and diligence," Boxton proclaimed. "My coaching has nothing to do with how good or bad the athletes are. What matters is how determined and diligent they are. They must be fierce and determined and set their hearts on being the best. Then, they must be diligent, doing whatever it takes through hard work and dedication to be the best."

"Are they the best yet, Coach?" Blaine asked.

"Not yet," Boxton replied. "And that's a bit confusing to me, because these athletes have shown me great amounts of determination and diligence, but they still didn't perform as well as I expected at the last Olympic trials or games. We have recently been focusing our attention on the health of the athlete's circulatory systems. Because, you see, science is just as important in athletics as knowing the rules to the game. The circulatory system is very important to these decathletes because it is responsible for transporting substances like food and oxygen around the body and for collecting waste substances for later disposal. It has three main parts: the blood, blood vessels, and the heart."

"How did it go with that doctor that contacted you?" Jek asked the coach. "The one that said he could help you increase the efficiency of the athletes' circulatory systems."

Coach Boxton's face turned red with anger just thinking about the dirty doctor.

"Dr. Veeginburger," the coach spat out in disgust. "It turns out, he was really no doctor at all. He just wanted to sell me illegal supplements that he promised would enhance our decathletes' performance. I showed him the curb as soon as he made his real intentions known. He wanted to put those nasty chemicals in the bloodstreams of our athletes, but chemicals can't replace determination and diligence. Plus, I have heard that supplements like Dr. Veeginburger's can have very negative effects on the heart. I will not put the hearts of any of my athletes in jeopardy."

Coach Boxton's integrity, plus his love for his athletes, was very inspiring to the Sassafrases. If they ever decided to train for a decathlon, they wanted him to be their coach.

"Here is an interesting fact," related Jek. "The ancient Greeks, who just so happened to be the first Olympians, believed that the heart was the seat of love and intelligence."

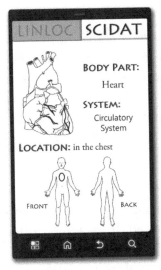

BODY PART:
Heart

SYSTEM:
Circulatory System

LOCATION: in the chest

FRONT

BACK

Coach Boxton laughed at this information. "Well, I don't know about love or intelligence, but I do know that the heart is responsible for pumping the blood that supplies the substances your body needs around the body. Arteries carry blood away from the heart, and veins carry it back to the heart. Both of them branch into tiny vessels called capillaries, which is where the exchange of substances actually takes place. The heart has two sides, known as the left side and the right side. Each side has two chambers called the atrium and the ventricle. The right side of the heart receives blood from the body and pumps blood to the lungs. The left side of the heart, which is larger, receives blood from the lungs and sends it out to the body. The atriums are responsible for taking blood into the heart, while the ventricles pump it back out. There are valves separating the opening and exit into the ventricles. This allows blood to flow into the chamber but prevents blood from flowing back to where it came from."

As the coach was talking about the heart, Tracey was entering the SCIDAT data into her phone and thinking back to something Boxton had said to Dr. Veeginburger while in the rickshaw about the blood being part of an intricately working machine. The more the coach revealed about the heart and the blood, the more that metaphor made sense. Honestly, Tracey took the heart for granted, not thinking about it most of the time, but wow! It really was like an amazing machine!

"The heart beats over two billion times in a lifetime, without stopping," Coach Boxton kept going. "Each heartbeat has three stages. First, the heart relaxes and brings blood into it. Second, the atriums contract, which pushes blood into the ventricles. Third, the ventricles contract pushing blood to the body and lungs. It is because of these stages that the heartbeat has a lub-dub sound."

Blaine reached up and felt his own heartbeat. From their spot high up in the bleachers, he looked down at the athletes on the field and thought about their heartbeats and the blood running through their bodies, enabling them to do what they were doing.

"Well, do you guys want to go down and meet the decathletes?" Coach Boxton questioned.

"Sure!" the twins responded in unison.

On the way down, the coach explained to the twelve-year-olds a little bit about how the decathlon worked. It was a competition comprised of ten separate events that took place over the course of two days. Boxton bragged about how healthy and versatile his athletes had to be to even be considered for the decathlon.

"They are the greatest complete athletes in the world!" the big coach declared, as they stepped out of the stands and onto the field.

As their esteemed coach approached, the Chinese decathletes, both men and women, all waved 'hello,' but then stayed focused on whatever discipline they were practicing at the moment. As the coach, Jek, and the twins walked around the track and through the middle of the field, Boxton continued to give details on the decathlon.

The twins found out that the five events on the first day were the 100-meter sprint, the long jump, the shot put, the high jump, and the 400-meter run. Then, on the second day, the decathletes had to complete the 110-meter hurdles, the discus throw, the pole vault, the javelin throw, and then, finally, the 1,500-meter run. These truly were impressive athletes. They seemed very self-motivated and very engaged in what they were doing.

Coach Boxton walked around and encouraged each of the athletes, with the always-smiling Jek translating for him. He also gave many of them helpful coaching points on how to perform their disciplines better, but when he returned to the front row of bleachers, where the Sassafrases were now sitting, he had a sizeable frown on his face.

"I just don't get it," he sighed. "These are the most driven athletes I have ever coached. I don't understand why we haven't gotten any Olympic medals with this group yet?"

The big coach plopped down on a seat next to the twins and let out a long deep breath. Jek, who was still standing on the track facing the three, had a thoughtful look on his face.

"Bu dao chang cheng, fei hao han," the teenager shared in Chinese.

"What does that mean?" Tracey asked.

Jek's smile returned, accompanied with some new and sudden excitement. "It means: 'One who has not climbed the Great Wall is not a real man.'"

"Oh, yes," Coach Boxton said. "I have heard this old Chinese saying before. It is interesting. Why is it getting you so excited, Jek?"

"Maybe this is the answer to your question about your athletes, Coach," Jek said enthusiastically.

"So what you're saying, is that if I take all the decathletes out to climb the Great Wall of China, they will become real men and women, and therefore, win some Olympic medals in the decathlon for China?" Boxton clarified, sounding skeptical.

"Maybe!" Jek grinned. "Maybe that is all the athletes need to reach their full potential. There is also a little known legend about climbing the Great Wall that could help us even more."

"Legend?" Boxton asked.

"Yes," Jek answered. He kept his smile, but he began talking lower and more slowly. "It is said that somewhere hidden within the

Great Wall itself is the Ancient Calligrapher."

"The Ancient Calligrapher?" Blaine asked.

"Yes," Jek confirmed again. "He is an old man that lives somewhere inside the wall. Legend has it that he possesses wisdom beyond compare, and that he has the ability to answer any question."

Now Coach Boxton was starting to get interested. "So I could ask him all kinds of questions about what my athletes need to do to get over the hump."

"Yes, but it is said that he will answer only one question," continued Jek. "And he doesn't utter any words when he gives his answer. He simply writes down the answer using age-old Chinese characters."

"So how do we find out where he is?" Tracey asked, enthralled.

"You see, that's the thing," the joyful translator responded. "His whereabouts have been a virtually complete mystery over the years. But, lucky for us, a member of my family recently came across some clues on how to find him. My uncle was one of the thousands of laborers that helped construct the very stadium that we are in right now, the Bird's Nest. One day, when he was cleaning up construction materials from a back hallway, he overheard the general contractor repeating a series of numbers and letters to one of the sub-contractors. He was telling the sub-contractor that these numbers and letters would be the connecting link to China's two greatest architectural marvels, one old and one new: the Great Wall and the Bird's Nest. The sub-contractor asked the general contractor how he knew this. He responded that the number and letter combination led to the answer, and the answer was given by the Ancient Calligrapher."

"My uncle scribbled down the numbers and letters on a broken piece of sheetrock and saved it. He is still convinced that those numbers and letters will somehow lead straight to the Ancient

Calligrapher's dwelling. But neither he nor anyone else in my family has figured out how."

"Do you have access to the numbers now?" Coach Boxton asked.

"I sure do," Jek responded. "Right here in my phone."

"Oh, c'mon!" the crazy, red headed scientist exclaimed. "Have a heart!"

He paused, trying to keep a straight face, then exploded in laughter. He patted Socrates on the back like he was a real person. Then he snapped a plastic heart inside the skeleton's frame.

"There you go, Sox," Cecil giggled happily. "Now you have your heart. But where is Aristotle's heart?"

He rummaged through the box of parts until he found the other heart that fit inside the second plastic skeleton. He snapped it into place.

"Now, stickers…stickers…where are the stickers? Oh wait—silly me. We don't need any more stickers until we get to the immune system."

Cecil stepped back and looked across Socrates and Aristotle as they stood now. He was adding pieces at the same pace as the twins were progressing through their locations. Most SCIDAT topics his niece and nephew were studying could be represented by these plastic snap-in pieces. But when a part of the anatomy wasn't easily represented by a plastic model, Cecil had stickers that he would stick to Socrates and Aristotle.

Cecil, even now, was looking at the stickers he had carefully placed to represent the five senses. So far, the two plastic skeletons stood there with only their skeletal, respiratory, and circulatory systems intact. His two plastic buddies were not complete yet, but they were looking good.

Cecil smiled. The scientist was confident they would be completed soon. He was confident in that because he was confident in Blaine and Tracey.

CHAPTER 9: THE ANCIENT CALLIGRAPHER

Muscle Men

Tracey had been quiet as they listened to Jek read the letter and number combinations. Then, suddenly, the answer hit her like a ton of bricks.

"It's the seats!" she blurted out. "The combinations you just read are referring to seat numbers!"

The three guys looked at her with speechless confusion, which slowly turned into understanding.

"Tracey! I believe you are correct," Coach Boxton exclaimed. "That has to be it. Can you read the first one again, Jek?"

Jek looked back at the screen of his phone and read the first number and letter combination, "G38."

"G38," Coach Boxton repeated. "Well, let's go find that seat and see if Tracey is right."

The four hopeful clue-seekers began walking up and through the rows and rows of bleachers, looking for the correct seat.

"Here is row G," Jek shouted out. "Now we just have to find seat number thirty-eight."

Blaine made it to the seat first and started looking it over for anything that stood out. There wasn't anything on the back of the seat or the arm-rests or on the seat top itself. The boy didn't see anything at all. Of course, he didn't really know what he was looking for. The other three joined him and started searching over the chair as well.

Once again, it was Tracey who figured it out first.

She ducked down and looked under the seat. There, etched on its underside, were several Chinese characters.

"Here!" the girl pointed out. "There is something written in Chinese on the bottom of the chair."

Jek hunched down and took a look at what Tracey had found. "Kai shi zai mu tian yu de pao tai," the boy read aloud.

"Is it a clue?" Coach Boxton asked.

"It's definitely a clue," Jek responded. "It means: 'Start at Mutianyu battlement.'"

"Mutianyu battlement?" Blaine questioned. "What and where is that?"

"A battlement is like a little fort built on the wall, and Mutianyu is a distinct section of the wall," Jek answered, getting more excited. "I know just where it is! It's only a few hours from here by bus. This is where we should start! It looks like my uncle was right! The numbers and letters that he heard the general contractor repeating really were clues, which means that the Ancient Calligrapher must be real too!"

"Well, let's not get too excited," Coach Boxton cautioned.

"There are still a lot more letter and number combinations to find on the seats in here. Let's see if we can find all the clues, and then we'll know better what to do."

Despite the coach's caution, the twins could tell he was getting more and more excited, and so were they.

Jek looked at his phone for the next combination, "Y49."

Everyone now went in search of the Y section of the stadium to look for seat number forty-nine. Jek got there first, closely followed by the others. He ducked down and looked under the seat. Sure enough, there was another clue, which he read aloud.

"Ba shi ba dong mian." He immediately translated it into English. "Eighty-eight to the east."

All four got puzzled looks on their faces as Jek repeated, "Eighty-eight what to the east?"

None of them had any idea what it meant, but that did not dampen their desire to look for the rest of the clues here in the Bird's Nest. Jek continued to read the combinations from his phone as the group continued to dash around the arena looking for the clues on the seats. By the time they found the last seat, they had forty clues in all, but only the first clue had the name of a place. The rest simply had a number and a direction.

Excited but tired after their clue-hunting, the four took seats on Q six, seven, eight, and nine. Coach Boxton looked like he was just about to say something, but instead, his face started turning red as he stared at one of the stadium entrances. Blaine, Tracey, and Jek all looked to see what the coach was glaring at and what he was so angry about. Then, they saw; there he was again. The mousy Dr. Veeginburger was walking into the Bird's Nest, and he wasn't alone. Following along was a huge tower of a muscle-bound man walking in with him.

"I told him that I never wanted to see him again," Coach Boxton growled, as he stood to his feet. "I cannot let him near my athletes!"

The coach hurried towards the aisle and rushed down the stairs two at a time, as he made a bee-line straight toward the doctor and his monstrous friend. Blaine and Tracey gave each other a nervous look, wondering what was about to happen. They knew the coach could easily handle the tiny doctor, but what about the big, muscular guy with him? The Sassafrases had never seen anyone so big.

They knew Coach Boxton was angry, and he had every right to be. They just hoped he would be careful. Jek and the twins got up and followed the coach down to the field. When they reached the track, they could hear that Boxton was already headlong into a confrontation with the shady doctor.

"I told you plainly back in the rickshaw, Dr. Veeginburger, we don't want and will never want any of your illegal supplements!"

"I know, I know," the little doctor sneered. "I know what you said in the rickshaw, but I just wanted to introduce you to Itsy and see if you might change your mind."

"Why would Itsy, here, change my mind?" Coach Boxton retorted.

Dr. Veeginburger laughed, "Well, just look at him, coach. He's huge."

Boxton didn't look impressed.

"And he's not just huge," Veeginburger continued. "He is also strong, fast, and agile. You want to know why, coach? It's all because he injects my supplements into his bloodstream every week. Just think. Your decathletes could have muscles that look like this."

Itsy took a step forward and flexed his muscles. At the sight of this, the twins' mouths dropped open. Itsy had looked large from high up in the bleachers, but now that they were standing up close, looking at his flexed muscles, he looked even bigger. His muscles seemed to ripple like waves on a pond, but Coach Boxton remained unimpressed.

"I said 'no'," he snarled through gritted teeth.

"At least let us demonstrate," Dr. Veeginburger pleaded, as he turned to Itsy and commanded, "Go. Run. Jump. Throw."

Itsy stopped flexing and immediately took off down the track. He was definitely fast, thought the twins, as they watched the big fellow take the first turn. All the Chinese decathletes that had been practicing stopped what they were doing and watched the demonstration. Itsy rounded the second turn and quickly approached a row of hurdles that had been set up on the track, but instead of jumping over them, he just ran right through them, smashing the hurdles to pieces as he went. An embarrassed look came over Dr. Veeginburger's face, but he quickly covered it with a nervous laugh.

Coach Boxton just folded his arms and shook his head. Itsy made his way around the last two turns and sprinted back to where the group was standing. He came to a stop with a big goofy grin on his face.

"Itsy, you big forgetful mess!" Veeginburger chided. "You ran, but you didn't jump or throw."

Itsy's grin turned to a frown.

"Go do it again," the mouse-like doctor demanded. "This time, show the coach how you can jump and throw."

"No!" Boxton exclaimed forcefully. "There will be no more demonstrations. You two are going to leave right now."

Dr. Veeginburger looked up at the coach arrogantly, "Who exactly is going to make us leave, coach? You? Itsy is twice your size. He could use those muscles of his to pound you into the ground if you try to make us leave."

Coach Boxton's response to the doctor's latest remark did not consist of words, but rather actions. He walked over to Itsy, grabbed the giant man, and flung him to the ground with some kind of judo toss. He then got down and cradled the big man's head and one of his arms together in a tight headlock. The twins could not believe what they had just seen and evidently, neither could Dr. Veeginburger, as all four stood there with stunned speechlessness.

"Let me tell you a little something about muscles," Boxton urged, still holding the big man in a headlock. "The muscular system helps the body to move, it supports the body, and it assists the body with vital functions, such as the beating of the heart. But bigger muscles don't necessarily mean stronger muscles, and drug-enhanced muscles definitely don't mean better muscles."

Now, Dr. Veeginburger's face was getting red as he shouted, "Itsy! Get up, you buffoon! You're bigger than the coach. Just pick him up and throw him off! Use those muscles that I have perfected for you!"

Itsy just looked blankly at the doctor, incapable of movement because of Coach Boxton's headlock.

"There are three types of muscles," the coach went on, "smooth muscles, cardiac muscles, and skeletal muscles."

"I don't care what you know about muscles," the little doctor shouted. "I just wanted to sell you my medicine!"

Boxton ignored the doctor and went on. "Smooth muscles help the body carry out normal functions such as digestion. They are found on the walls of hollow organs, such as the stomach and intestines. They contract smoothly and rhythmically. The smooth muscles are controlled involuntarily, which means that most of the time we are not even aware of their movement. The

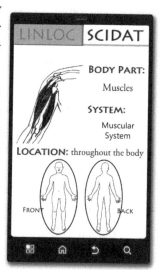

cardiac muscle is found only in the heart, and it helps the body maintain its heartbeat. It contracts continuously and is under involuntary control so we don't have to think about making our heartbeat. Finally, my skeletal muscles are helping me to keep little Itsy here pinned down! They—"

"I don't care!" cried Veeginburger. "I don't care what you know about muscles. I now know you are a better scientist than I am. I also know that you will never in a million years buy my supplements. Just let Itsy go and let us out of here!"

Coach Boxton looked at the sniveling doctor. "If I let him up and let you leave, are you going to come back? Are you ever going to threaten the health of my athletes again?"

"No, never," moaned Veeginburger. "I will never even think about your athletes. You will never see me again!"

Boxton gave Veeginburger a long serious stare, as if contemplating the doctor's sincerity. He then quickly released Itsy from the headlock, hopped up, and helped the big guy back to his feet. Without saying a word, Boxton held out his arm and pointed toward the exit. Both Dr. Veeginburger and Itsy wasted no time as they made their way quickly out of the stadium. As soon as they disappeared, Coach Boxton's glare turned to a smile, and he clapped his hands together a couple of times as if cleaning them off.

"Well, it's getting late," he said to the three. "Let's go inform the decathletes that practice is over for the day and that tomorrow we will be taking them on a field trip to the Great Wall!"

He gritted his teeth in disgust. He had been so close to catching them, but he hadn't been able to pull it off. He had followed the twins to the Bird's Nest and had been able to hide himself up in the bleachers. When the two of them began wandering all over the stadium with that coach and his translator, looking for specific chairs, he had found himself just three rows up from where the Sassafrases were.

He had his expandable trap boxes, one in each hand, and he was just about to throw them at the twins, when the coach had suddenly shouted out that he had found the next chair. The twins had immediately darted over to where the coach was, and out of his throwing range. It was another opportunity missed.

However, he happened to know that the twins and their friends were planning to take a trip to the Great Wall in the morning. Maybe he could find a way to catch them in his boxes there.

Muscular Contractions

The next day, after a good night of sleep, some picture-selecting, and scientific data entry, Blaine and Tracey climbed on a privately chartered bus with Coach Boxton, Jek, and all the Chinese decathletes. The driver fired up the big bus and off they went, headed out of the city toward the Great Wall of China. The Sassafras twins were excited. They were excited about seeing the wall, and they were excited about the growing archive of science that they were amassing. Even though school had let out just a couple of weeks ago, the memories of how much they disliked science were quickly fading.

"Wow," Blaine exclaimed, after an hour or so on the bus, "there it is!"

Tracey looked out of the window her brother was gazing through.

"Wow," she agreed, sharing Blaine's sentiment. "It's really here. I've seen it in pictures, but now there it is, the real thing: the Great Wall of China. Look how it snakes over the mountains and valleys. Look how cool its forts, or what did Jek call them, battlements? Look how cool they are!"

"Yeah," Blaine nodded, "and it's just...great...it's a great wall...it's the Great Wall."

The driver found a suitable place, parked, and everyone filed off the bus. Coach Boxton looked at all of his athletes. They really were China's best.

"One who has not climbed the Great Wall is not a real man," he stated.

"Bu dao Chang Cheng, fei hao han," Jek translated.

Many of the athletes started smiling. They were already catching their coach's drift. He was trying to inspire them. He was reaching into their culture, pulling out something to awaken their drive, and creating an image that would fuel their determination and diligence.

One of the male decathletes that the twins had seen throwing the shot put the day before, raised his fist in the air and shouted something in Chinese. Whatever he said got all the other athletes excited. They, too, raised their fists and passionately shouted the same thing. Then, following the first athlete's lead, they took off up the hill as a group toward the Great Wall, as Coach Boxton smiled proudly.

"See how great they are?!" he declared to the twins. "There is no way I would ever pollute their bloodstreams or jeopardize their hearts or muscular systems with any of Dr. Veeginburger's dirty supplements!"

The decathletes reached the wall in no time. They climbed up some stairs and entered one of the wall's battlements. Then, as a united and determined group, they started a run along the Great Wall.

"Just watch them go," Boxton said proudly. "They have the drive and the brains to be Olympic champions. They have conditioned and prepared their muscles for competition. Did you know that muscles are made up of microfilaments that contract when told to do so by the brain? It starts when the muscles receive a message from the brain through the nerves. Next, the filaments of actin and myosin slide past each other and cause the fiber to get shorter, which results in the muscle contracting. All this happens lightning fast, so that we don't realize how much is really going on when our muscles contract."

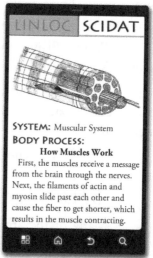

SYSTEM: Muscular System
BODY PROCESS:
How Muscles Work
First, the muscles receive a message from the brain through the nerves. Next, the filaments of actin and myosin slide past each other and cause the fiber to get shorter, which results in the muscle contracting.

"What about the skeletal muscles?" Blaine asked. "You never did get to finish telling us about those yesterday."

"Ah, yes, the skeletal muscles," Coach Boxton recalled. "They help the body move and support the body when we sit or stand. They always work in pairs around a joint because they can only pull in one direction. For instance, when you want to move your arm, either your bicep contracts, bending your arm at the elbow joint, or your triceps contracts, causing your arm to straighten back out again."Tracey bent her arm at the elbow to observe what the coach was talking about.

"Skeletal muscles are the most common type of muscle found in the body," the big coach continued. "They are controlled voluntarily, except in the case of reflexes. So, for the most part, we know when they are going to jump into action. Skeletal muscles are joined to the bones by tough cords called tendons, enabling them to pull on the bone as they cause the body to move."

Blaine moved the fingers on his hand from open to a fist, thinking about the tendons running through his hand and the muscles in his arm working together to pull on his finger bones.

"Skeletal muscles are made up of bonded cylindrical cells called fibers," said Coach Boxton. "Inside each of these fibers are the filaments of actin and myosin, which I mentioned a second ago, along with one nucleus and huge numbers of mitochondria, which produce the energy for the muscle fiber to contract. These fibers are organized into bundles, and the bundles are wrapped together inside a tough sheath. All these bundles form the muscle. Pretty cool, huh?"

The twins nodded that it was indeed cool.

"What I don't understand," the Coach replied, with an annoyed look on his face, "is how these Chinese athletes can have the muscles, the blood, the hearts, and the drive, but no Olympic medals in the decathlon?"

"Well, hopefully we are about to answer that question," Jek smiled. "We have the sequential clues that we found on the chairs in the Bird's Nest. We are at the Mutianyu section of the Great Wall. We can see the battlement that we need to start at. Coach Boxton, we are all set to find the Ancient Calligrapher!"

The big coach smiled and patted his friend and translator on the back.

"OK, then, Jek. Let's do it. Let's go and see if we can find this guy."

Coach Boxton, Jek, Blaine, and Tracey all started up the hill toward the Great Wall and the first battlement. There were a few tourists in the area, spread out here and there, taking pictures or hiking. For the most part, this was a very lonely and quiet section of the wall, which was probably a good thing, considering they were looking for the secret entrance to the dwelling of a legendary folk hero. They reached the wall and climbed the stairs entering the battlement. Jek pulled out his phone and read the first clue they'd found at the Bird's Nest.

"Kai shi zai Mutianyu de pao tai. Start at the Mutianyu

battlement. Okay, so that is where we are now," Jek affirmed.

"Eighty-eight to the east was what the next clue said, right?" Tracey asked.

"Right, but we never did figure out what the eighty-eight meant," Jek answered, looking at his phone.

"Surely it doesn't mean go eighty-eight battlements to the east?" Boxton inquired. "That would just be too far."

"Maybe it just means eighty-eight steps," Tracey offered.

"It could mean that," Jek affirmed. "Then again, different people take different sized steps."

Everyone nodded their heads, as if agreeing.

"What about these?" Blaine said, speaking up from just outside the east door of the battlement. He was pointing to one of the cut-outs in the pattern on the defensive north side of the wall.

Jek's eyes immediately lit up. "Yes! Blaine, I think you might be right! Those are called defensive gaps, and I bet we are supposed to count out eighty-eight of those, walking east on the wall!"

Blaine and Tracey joined Jek in his excitement. Coach Boxton was excited too, but he was also still a little skeptical.

"Blaine may or may not be right," the big coach admitted. "It is definitely worth a try, counting these defensive gaps out. If we are going to do this, though, we'd better get started because we have a lot of clues to cover, and a fair amount of those clues have big numbers. So we could be in for quite a bit of walking and counting."

Starting from the first defensive gap on the east side of the first Mutianyu battlement, they began their count of gaps. Blaine had been thinking out loud when he'd wondered if the wall's defensive gaps were what they were supposed to count. He sure hoped he was right. If not, they were going to be doing a whole lot of counting for nothing. Also, they had to be very careful not to lose count. They didn't want to have to start all over.

Counting out eighty-eight gaps proved to be easy enough for the group. When they reached the last gap, Jek read the next clue, "Three hundred forty-five to the west."

Now the group headed back in the direction they had just come from, counting gaps as they headed west. They went back and forth like this, along the wall, going up and down stairs, traversing flat spots and steep spots, passing through battlements and forts.

When they were about halfway through their clues, the Sassafrases heard the voice of a passing tourist say something that jogged a not-too-distant memory. "This is totally sick, guys! I can't believe we're actually trekking the Great Wall of China!"

Both Blaine and Tracey lost their count and froze in their tracks. They knew that voice. They turned their heads and saw the face of the person that had uttered the memory-jolting sentence.

"Gretchen!" they blurted out.

The pretty brunette, earring-clad backpacker girl stopped and looked at the twins with a look on her face that said, "How do you know my name?" But that look slowly turned into joyful recognition.

"Blaine and Tracey?" Gretchen questioned. "Blaine and Tracey Sassafras?"

The twins nodded with beaming smiles. Of course, Gretchen wasn't alone. She was accompanied by her two gnarly trekking buddies, Skip and Gannon.

"Dude! It's the Sassafras twins," exclaimed the mohawked Skip.

"Man, it is crazy to bump into you guys here on the Great Wall, man," Gannon gushed, while bobbing his head a little, causing his dreadlocks to bounce.

"It is crazy," the twins thought. They had just seen these

three not even two weeks ago in Peru, while studying zoology. How had the three trekkers gotten all the way to China so fast? Then a heavy thought hit the twins: "Were Skip, Gannon, and Gretchen using invisible zip lines to travel the world as well?"

Blaine impulsively asked, "How did you three get here? Do you guys have invisible zip lines too?"

Tracey elbowed her brother in the ribs. Skip, Gannon, and Gretchen all responded with blank looks on their faces. "Invisible zip lines?" Gretchen repeated. "That would be sick, wouldn't it?"

"Dude, that would be rad!" Skip added.

The three trekkers laughed, thinking that Blaine's question had been some sort of joke.

Gannon looked at Blaine. "No, man, we didn't get here by invisible zip lines. We rode on a plane, man."

The three trekkers laughed some more, as did Coach Boxton and Jek, who had heard Blaine's question as well. The Sassafrases both nervously joined in the friendly laughter, hoping their secret was still safe.

"Dude, we are just traveling the world like the two of you are," Skip explained. "We did Peru, and now we're doing China. It sure is a small world, dude."

"Yeah, it is sick bumping into you two here on the Great Wall," Gretchen agreed.

After a little more small talk and introductions, the Sassafrases said goodbyes to Skip, Gannon, and Gretchen and continued on in their count of defensive gaps. The twins had forgotten what number they were on, but luckily both Coach Boxton and Jek had remembered.

The day wore on and their excitement grew as the group of four continued working through their clues. The twins thought not only about the Ancient Calligrapher but also about how amazing the Chinese decathletes were for running this section of the Great Wall. It wasn't flat and easy; rather, it was steep and difficult. Their respect for these amazing athletes was expanding.

Finally, after a couple hours of counting and walking, the four got to their final clue. "One hundred and eleven to the east," Jek read aloud.

The teenager then let out a long, hopeful, deep breath. "I sure hope we have kept the correct count!"

The others agreed as they began counting out their last clue. One hundred and eleven defensive gaps brought them right up against the outside wall of a battlement. They had already passed through this particular battlement a couple of times as they had traced back and forth, following the clues' numbers. None of them had noticed anything out of the ordinary anywhere they'd been so far on the wall, and now that they were at the end of the clues, they still didn't see anything unusual.

"What are we looking for now, Jek?" Coach Boxton asked. "How do we know if we've arrived at the right place?"

Jek just shrugged like he didn't know. "My uncle wrote down all the numbers and letters that he heard the general contractor repeating. We found all the clues corresponding with those combinations on the seats in the Bird's Nest. We have counted out all of those here on the Great Wall, and this is where we have found ourselves, standing against the wall of this battlement. The Ancient Calligrapher's dwelling just has to be here somewhere."

"What's this?" Tracey asked, rubbing her fingers on the wall over some Chinese characters she'd spotted.

Jek stepped over to where she was and studied the characters. The biggest smile the twins had seen yet appeared on his face.

"What does it say?" Coach Boxton questioned, excitedly.

"It says: 'Bu dao pao tai, wu da an', which means: 'One who hasn't climbed this battlement is not an answered man,'" Jek translated.

"This has to be it then, right?" Tracey almost shouted. "This has to be where the Ancient Calligrapher is!"

"I bet it is," Coach Boxton grinned. "'One who hasn't climbed this battlement is not an answered man' suggests that if you do climb it, you will be an answered man, and that is what the Ancient Calligrapher does, right? He answers questions."

No one had to persuade Blaine to believe this was the spot they had been looking for. He had found a place on the battlement's wall that was possible to climb, and he was already half-way up. The other three happily followed his lead and began the climb up behind him.

Once on top, they found that the only thing that was there was a solitary Chinese flag, flapping in the wind at the top of a wooden flagpole. The four scanned their eyes across the top of the battlement, looking for another message written in Chinese characters or any other possible clues. None of them spotted anything as they pulled and pushed on different stones. They traced their fingers across every gap and crevice, but nothing stood out. Nothing moved.

"Maybe we have just been on a wild goose chase," Jek suggested sadly. "Maybe the Ancient Calligrapher is just a legend. Maybe he doesn't really exist."

On a whim, Blaine walked over and grabbed the flagpole with his hands and twisted it. Suddenly, to everyone's surprise, the flagpole dropped about a foot down into the stone roof of the battlement. A grinding noise could be heard as part of the surface began to slowly drop away, creating a stone staircase that led down into darkness.

Coach Boxton, Jek, and the twins all had half joyful and half unbelieving looks on their faces. Had they really just found the dwelling of the legendary Ancient Calligrapher?

Each of them took the stone steps slowly, one step at a time, more out of awe than timidity. The staircase took just one turn and then led them straight down into a small dark room. It took their eyes a moment to adjust to the lack of light, but when they did, they saw him.

Blaine and Tracey gasped. There sat the Ancient Calligrapher, hidden right there in his secret room inside the Great Wall of China. He was a very small man with a face full of wrinkles. He had a bald head and a long whispy white beard. He was sitting with his legs crossed, completely silent and still on the stone floor.

"Hello," Coach Boxton said respectfully. "We have come here to ask you a question."

Jek translated, but for a long while the old man just sat quietly, without a movement. The twins even began to wonder if he was a statue and not a real person. Coach Boxton was about to repeat his statement when the Calligrapher slowly reached out and lit a candle none of them had even known was there. The dim, flickering light of the candle revealed three items laying on the floor in front of the old man: an old piece of rough parchment, a slender paintbrush, and a wooden bowl filled with black ink.

Coach Boxton looked at Jek and whispered quietly, "Does this mean I can ask him a question now?"

Jek shrugged. "I guess so, Coach."

"Just one question, right?" Boxton asked.

"That's what the legend says," Jek confirmed. "But I don't know if that means one question per person, one question per day, one question per visit, or what."

Coach Boxton nodded in understanding before he looked back at the ancient man sitting on the floor.

He cleared his throat and then inquired, "What will it take for my decathletes to become Olympic champions?"

Jek slowly repeated the question in Chinese. All four then waited with bated breath to see if the Calligrapher would answer. Again, the old man sat completely silent and still for a long time. None of them dared to prod him along or even make a movement or sound.

After what seemed like forever, the Ancient Calligrapher slowly reached down and grabbed the brush. Every movement he made was done precisely and at a snail's pace. He held the paintbrush perfectly vertical as he carefully dipped it into the bowl of black ink. He brought the brush up out of the ink and then gingerly placed its bristles down on the parchment. Then, with the skill of a gifted artist, he delicately painted out each and every stroke of every character. The movement of his hand, as he composed the strokes of the characters, seemed to be second nature to him.

When he was finished writing down the answer to Coach Boxton's question, he put the brush down, picked up the parchment, and held it out with both hands.

Coach Boxton respectfully stepped forward, thanked the man, and carefully took the answer from the Ancient Calligrapher. Boxton looked over the beautiful calligraphy that had just been inscribed, but he obviously couldn't read it. So he handed the parchment to Jek.

The Chinese boy took the old rough piece of paper and read the Ancient Calligrapher's answer. "Jue xin he qin fen." Jek smiled and immediately started laughing.

"What's so funny?" Coach Boxton asked. "What does it say?"

"Your question was," Jek recalled, "'What will it take for my decathletes to become Olympic champions?'"

The coach nodded at the restatement of his question.

"The Ancient Calligrapher's answer is..." Jek paused dramatically. "Determination and diligence."

For a split second the coach's face was covered in bewilderment, but then he joined Jek in laughter.

"How ironic?" thought the twins. "Coach Boxton already knew the answer to his question." As the twins watched him happily laughing with Jek right now, they could tell the Ancient Calligrapher's answer had in no way discouraged him. Instead, it had greatly encouraged him. Probably because he knew he just needed to keep doing what he had been doing; to keep coaching how he had been coaching; to hone in on the importance of determination and diligence that would get the Chinese decathletes an Olympic medal.

A sudden realization hit Blaine right in the gut. He had a question that he desperately wanted to know the answer to. Something had been gnawing at his mind day after day. Without even passing the idea by Tracey or asking Jek if he thought it was okay, the Sassafras boy uttered his question out loud in the direction of the Ancient Calligrapher.

"Who is the Man with No Eyebrows?"

CHAPTER 10: TRAPPED IN TEXAS!

Digestion starts with the teeth...

The question hung in the air for just a second and then, suddenly, the Ancient Calligrapher's candle blew out and the stone staircase started closing up.

"Evidently this means he is not going to answer a second question today," Coach Boxton declared. He then looked over at the closing staircase in urgency. "Our exit is shrinking! We have to get out of here fast!"

All four bolted like streaks of lightning for the staircase. The steps seemed to be closing a lot faster than they had opened up, but they all managed to make it back out to the roof in plenty of time before the stone steps once again became a part of the flat surface on top of the battlement. The coach, the translator, and the twins all just stood silently for a moment, amazed at what they'd just experienced.

It was Boxton that broke the silence. "I'm sorry you didn't get your question answered, Blaine."

"Oh, that's all right, Coach," the Sassafras boy sighed. "I'm just glad we figured out the clues and that you got an answer from the Ancient Calligrapher."

Coach Boxton smiled like he still couldn't believe it was true. He held up the parchment on which the Ancient Calligrapher had inscribed his answer.

"Determination and diligence," the coach repeated. "Now I know what to do: just keep on keepin' on."

Later that evening, as Coach Boxton and Jek were tracking down their amazing Chinese decathletes who had basically been running all day long, Blaine and Tracey found a secluded spot there on the Great Wall and finished entering their SCIDAT data. Once they had added their archived pictures and pressed send, they moved on to the always-exciting opening up of the LINLOC app to see where they would be going next.

"Texas?" Tracey exclaimed in question form, as she read the next location. "Lubbock, Texas? We are going from Ethiopia, Australia, Italy, and China to Texas? Interesting," the Sassafras girl stated.

Blaine looked at his sister with a sly grin. "Texas—it's like a whole other country."

This was his chance! The twins were finally alone, sitting there preoccupied with their smartphones. If he was going to catch them in his Expandable Trap Boxes, it had to happen now. He began

THE SASSAFRAS SCIENCE ADVENTURES

loosening up his throwing arm.

LINLOC SCIDAT

LOCATION: Lubbock, Texas
CONTACT: Burly Scav
LATITUDE: 33° 34'N   **LONGITUDE:** -101° 51'W

INFORMATION NEEDED ON:
Digestive and Urinary Systems

"Burly Scav," Blaine informed. "That's the name of our next local expert. It looks like we will be gathering information on the digestive and the urinary systems."

"In Lubbock, Texas," Tracey repeated the location. "Longitude 33° 34' N, Latitude -101° 51' W."

Oh, no! A vibration of angst shot down his spine. They'd already opened up the LINLOC application! That meant they were about to leave! He had to trap them in the boxes before they zipped off to their next location. He may not have another opportunity as good as this one.

Both twins put on their helmets. It was time to zip. They cinched up their harnesses tight and then turned the rings on their carabiners to the coordinates that would send them to Texas.

He tried to calm down his shaking hands. If his aim wasn't accurate, this wouldn't work. He had to throw the boxes in direct lines right toward each twin. The small boxes would then expand in the air and land on the twins, trapping them inside. The boxes, when expanded and closed, were unopenable, soundproof, and, most importantly, they could block any and all satellite signals. This, of course, would make it impossible to send or receive any texts or phone calls. It would be the end of the Sassafras twins. His revenge on Cecil Sassafras was about to take flight.

The coordinates were in and their carabiners had found the correct invisible zip lines. Each twin was now hanging with their feet dangling in the air. The twins smiled at each other. This was always a fun and heart-pounding time. The few anticipatory seconds they had just hanging in the air in between their carabiners connecting to the lines and zipping off.

Both of his arms went forcefully forward as he threw an Expandable Trap Box from each hand. They sped directly toward the Sassafras twins. His aim looked to be good, and he watched the boxes expand in flight.

There was nothing like it. Blaine and Tracey wiggled their fingers and toes in anticipation. Hey, what was that coming toward them? Smash. Crash. Darkness.

THE SASSAFRAS SCIENCE ADVENTURES

Yes! It worked! He had done it! His aim had been true. The boxes had expanded and trapped like they were supposed to. He had finally caught the Sassafras twins. Now he just needed to figure out a way to haul them off. He started walking towards them when all of a sudden the two boxes....disappeared.

Wait, how was that possible? He had trapped them, hadn't he? He reached up and wiped some sweat from his eyebrow-less forehead in dismay. Evidently, the invisible zip lines still worked, even if they were shut inside of an expandable trap box.

Tracey was stunned. What had just happened? She now felt like she was zip lining at light speed, but there was no light. Only darkness. What had happened? Who had turned out the lights?

Blaine blinked his eyes, expecting to be able to see something, but there was nothing. He couldn't see a thing. He felt the familiar feeling of his body cruising over the zip lines, but shouldn't there be light? The zip lining feeling continued for a few seconds and then Blaine felt his body come to a jerking stop. He must have been on the invisible zip lines, and he must have just landed, but still there was no light.

He had reached out with his hands and feet in every direction. In every direction, all Blaine felt was shut, blocked, and solid. The Sassafras boy slid his backpack off, unzipped it, and pulled out his phone. He turned the smart phone on and started dialing his uncle's number. Then he saw something so disheartening he almost couldn't handle it. The blue glow from his screen confirmed his suspicions. He was trapped in some sort of box.

Where was Tracey? Was his sister with him? "Tracey!" Blaine shouted.

Tracey recognized that. That was the jerky feeling at the end of a zip line journey, but she was still shrouded in darkness. The twelve-year-old girl remained completely still until her strength returned. Her strength was also usually accompanied by sight. Not this time. There was still nothing but darkness. Suddenly, Tracey felt very alone. What had happened? Had she been separated from Blaine again? She reached out her hand, and before her arm was even fully extended, her hand bumped into something hard. Whatever it was, it was flat and smooth…and solid.

Somewhere in a quiet neighborhood in Lubbock, Texas, very early in the morning, two rather large but nondescript boxes appeared on a curbside amongst a cluster of trashcans. Within moments of their arrival, a big trash truck turned onto their street. Driving the truck was a big, unshaven, pot-bellied man who was always smiling and continuously talking. Sitting beside him was his hoodie-wearing, rail thin teenage son, who seemed to be permanently frowning and virtually never speaking. They were the Scavs, Burly and Trevor Scav.

This fine morning found the father, Burly, chatting up his son about the good old days, and Trevor, the son, shutting his father's stories out by listening to loud rock-n-roll music on his headphones.

"…and that's when I learned never to drink milk before football practice!" Burly finished, smiling and laughing as he came to the end of his story.

Trevor hadn't heard a word of his dad's story, and he hadn't wanted to. He was just glad that they stopped the truck so often, and that he was the one who got to get out and dump the trash cans into the back of the truck. It gave him breaks from his dorky dad, however short those breaks might be.

They had just turned onto their last street of the morning. After gathering the contents of the cans on this street, they would head off to the landfill where they would dump the entirety of the truck's contents into a big hole in the ground. Burly stopped the truck, and Trevor immediately jumped out to tackle what looked like quite a big pile of trash cans. He grabbed the first can that he saw, took the lid off, and chucked the rubbish out of the can into the truck. He threw that can back down on the curb and reached for the second can, when he saw something he hadn't expected to see: two big boxes.

"That's weird," thought the teenager.

The sound had started out as a dull sound, but now it was loud, and Tracey thought she had figured out what it was.

"That's the sound of a truck," Tracey assumed. "It just has to be." Then, she heard the clanging of some metal around her like there was someone or something outside of the truck doing something. She had already tried to call Uncle Cecil, but her phone didn't have a signal. She had tried with all of her strength to pry open the box that she was now in, but nothing budged. Now it was time to yell.

"Help!" Tracey shouted at the top of her lungs. "Help me! I'm stuck inside this box!"

Trevor wasn't sure why someone would throw out two big boxes like this. He couldn't tell exactly what they were made of, but they looked sturdy and not at all like garbage. Trevor paused for a moment, adjusted his headphones, and stared at the strange boxes.

"Oh, well," he finally supposed. "Who am I to question what someone throws out?" The teenager picked up the heavy boxes one at a time and threw them into the trash truck.

Blaine had yelled. He had banged on the box with both his hands and feet, but whoever had just moved his box hadn't heard him. Was it possible for him to be able to hear things on the outside of the box but for those things on the outside of the box not be able to hear him on the inside of the box?

"This is bad," Blaine felt. "I'm stuck in a box. I'm separated from Tracey again, nobody can hear me in here, and I can't call Uncle Cecil. What on earth am I supposed to do now?"

Burly and Trevor finished dumping all the trash cans on the street and then off they went toward the landfill. Burly smiled as he drove the long, flat, straight roads of Lubbock. Today was a big day for him and he was excited about it. He had been a trash man for over twenty years, but he had always fancied himself as an inventor as well. Today, he was presenting to a potential investor a science project of sorts that he had been working on for a long time. It was something that he called Smart Dump.

Over the years, Burly had seen firsthand just how much space trash and garbage took up. The landfills were growing too fast, at least that's what Burly thought. So, he had designed a test site for a new kind of landfill. His premise was that a landfill could work much like the human body's digestive and urinary systems. Instead of dumping trash, as is, into landfills, which creates mountains and mountains of trash, it would be dumped into a Smart Dump site, where the trash would be compacted and could even produce energy or water.

The only person he had ever told his full idea to was his son, Trevor, but Trevor hadn't seemed too impressed. Trevor seemed to like his friends, his music, and really anything, better than his dear old dad. However, Burly loved his teenage son with an unconditional love—a love that couldn't be broken by any attitude or action from his son. Even though Trevor had been making some bad decisions lately, Burly was optimistic that his son was on the verge of a positive turnaround.

"Ahhhh!" Burly announced, as they pulled off the paved road onto a dirt road. "Here we are at the county dump."

Trevor just sighed, looking bored out of his mind.

They bounced down the dirt road toward the landfill, but instead of going to the main dumpsite, Burly guided the truck to where he had prepared his Smart Dump presentation. When they arrived, they saw there was already a vehicle there waiting for them. It was no ordinary vehicle—it was a black limousine.

"Whoa," Burly said, amazed. "Look, Trev! The Broadstine Group sent a limousine!"

Burly swung his trash truck around and then slowly backed it up to a sizeable hole in the ground. Then, the big man hopped out of the truck's cab and went bounding over toward the limousine. Before he reached it, the chauffeur got out and carefully opened the passenger door.

Out stepped a blond-headed woman, wearing heels and a skirt suit. She looked to be in her early twenties and not very happy to be there.

"Hello," Burly greeted her cheerfully. "I'm Burly Scav. Nice to meet you."

The big trash man held out his hand for a handshake. The young woman looked at Burly's hand in disgust. She then walked right past him, rolling her eyes.

"Just show me this Smart Dump thing," she huffed. "My father is forcing me to review this project, and I don't want to have to stand here one second longer than I have to."

Unfazed by the woman's rudeness, Burly smiled and walked back toward his trash truck. "What's your name, Ma'am?"

"Kimlee," the young woman barked. "Kimlee Broadstine."

"Well, then, Ms. Broadstine, right this way," Burly guided. "The show starts at the truck."

Tracey was completely baffled as to why they couldn't hear her yelling from inside this box. She could hear their voices as clear as day, but they, evidently, couldn't hear her shouts for help. They hadn't come to help her, and she had been yelling for a good long while. She had distinctly heard two voices — a man's and a woman's—but instead of coming to help her, the two had just continued their conversation.

"So, as much as possible, Smart Dump is designed to work like the human body's digestive and urinary systems," Burly was saying. "The digestive system, as you know, is responsible for breaking down food so that the body can use the nutrients. It consists of the teeth, the esophagus, stomach, small intestines, and large intestines. But, did you know that food goes through four distinct stages in the digestive system?"

Kimlee just rolled her eyes, and Burly pushed on, "The first stage is ingestion, where food is taken into the mouth, chewed, and then swallowed. Then comes digestion, which is where the food is broken down by muscular crushing and enzymes in the stomach. The third stage is absorption. In this phase, nutrients from the food are moved into the bloodstream through the intestines. The fourth and final stage is egestion, where waste is ejected out of the body."

"So where does the digestive system start?" Burly asked the young businesswoman.

Again, Kimlee rolled her eyes as she spat out, "I don't know, and I don't care. Just hurry through this presentation so I can get out of this disgusting place."

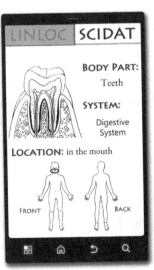

"That's right, at the teeth," Burly confirmed as if Kimlee had answered his question correctly. "Your teeth cut and crush food into small pieces, making it easier to swallow and digest. Teeth are covered with hard white enamel, but inside, there is an inner pulp cavity, which contains blood vessels and nerve endings. There are four main types of teeth. First are the incisors, which slide past each other and cut in the front part of the mouth. Second are the canines, which grip and pierce food. They are next to the incisors on either side. Third we have premolars, which are flat-topped and are used for crushing food. They are located between the molars and canines. And lastly there are the molars. They are larger, flat-topped teeth used from crushing food. They are found at the back of the mouth."

Burly paused to take a breath before he asked Kimlee another question, "And why do you think I told you all of this about teeth, Ms. Broadstine?"

Kimlee shook her head in annoyance, "I don't know why you told me all of that because this is a business proposal, not science class."

Not easily discouraged, Burly continued, "It's because I have rigged the inside of my trash truck with mechanical teeth that crush, pierce, and cut, just like the teeth in our mouths. Instead of just compacting the trash like most trucks do, my truck can use its mechanical teeth to break the waste down even smaller. Let me show you. All I have to do is push this button."

Inside the Stomach

Blaine had put two and two together. He was trapped inside some kind of box, and then that box had been picked up and thrown into a trash truck. The trash truck was equipped with mechanical teeth, and Burly Scav, their expert at this location was about to push a button to turn on those mechanical teeth. Blaine didn't know how it was possible from inside this box, but he had just heard everything Burly had said about teeth and the digestive system.

Blaine started yelling at the top of his lungs and beating on the inside of the box for all he was worth. He had to get out of here somehow. He had to get Burly's attention.

Using his thumb, Burly firmly pressed down on the button. Immediately, he heard his truck's mechanical teeth grinding to life.

The sound put a huge lump in the pit of Tracey's stomach. The mechanical teeth that she had just heard Burly talk about had turned on. "And now," she thought, imagining the worst, "those teeth are about to crunch up all the garbage in the truck, including the box that I am stuck in."

Both Burly and Kimlee peered into the truck and watched the metal teeth smash and chew up all the garbage. Kimlee was covering her nose and mouth with her hands, doing her best not to throw up. When Burly was convinced that the teeth had crunched up everything to a satisfactory level, he pushed the button again, stopping the mechanical teeth and pointed to a lever near the button.

"And this lever, Ms. Broadstine, is what I call the 'esophagus lever.' When I pull this lever, a chute folds out from the back of the truck. All the truck's contents are pushed out and carried through this chute, down into that hole you can see there in the ground. It's just like in the digestive system: After the teeth chew up food, the tongue pushes the chewed food to the back of the throat where it is swallowed. The epiglottis covers the windpipe as food enters the esophagus, which connects the mouth to the stomach. Its walls are smooth and slimy so that food easily passes. Muscles in the esophageal wall contract behind the food to push it downward. It takes about five seconds for your food to move from the throat into the stomach."

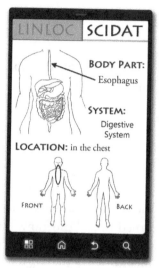

Burly stopped talking for a second and pulled the lever. All the chewed up garbage began to be pushed out of the trash truck into the chute and then began dropping down into the hole in the ground.

"Isn't it cool?" the trash man smiled. "Like I said, it works just like the body's digestive system—" Burly's sentence was interrupted by the loud growling sound of a nearby front end loader starting its engine.

"Oh, sorry, Ms. Broadstine," Burly shouted over the noise. "I was afraid this might happen. But don't worry—I have a back-up plan that will enable you to hear me."

The joyful trash man hopped toward the driver's side door of the trash truck and opened it up. He grabbed a small cardboard box out of the door's side pocket and raced back to where Kimlee Broadstine was still standing, holding her nose. Burly reached in the box and pulled out a small microphone, which he clipped on his dirty shirt near the collar. He then held the box out, which now only held earpieces for listening, offering them to Kimlee.

"Just stick one of these earpieces in," Burly advised over the noise, "and you will be able to hear every word I say."

But instead of reaching in the box and pulling out an earpiece, Kimlee took a big swing at the box and knocked it out of Burly's hand.

"I don't really want to hear anything you have to say, Bernie!

Just hurry up and finish so I can get out of here!" she screamed.

The cardboard box and all the earpieces flew up into the air and then came sprinkling down into the esophagus chute.

Blaine couldn't believe it. He himself and the box that he was in had somehow survived the trash truck's mechanical teeth. He could tell that the box had taken quite a beating because there were now small holes and cracks across the box. He could finally see some light! That was the good news, but the bad news was that he had just heard Burly say the trash was now going to get sent down something called the esophagus chute.

Tracey was relieved and terrified all at the same time. She and her box had withstood the mechanical teeth, and she could now see out of several holes created by those teeth, but even now, she could feel her box sliding downward at a fast rate of speed. She presumed she was on the chute that she had heard Burly mention.

Hey! What was that? Something had just fallen into her box through one of the holes.

What Kimlee saw was a big, stinky, dirty man in front of her that she wanted to be finished with. What Burly saw was a spoiled young woman that had a worse attitude than even Trevor, but he was determined to be kind to her.

What neither of them saw was that the earpieces Kimlee had knocked out of Burly's hand had not only landed on the esophagus chute, but they had landed on a couple of beat up boxes that were sliding down that chute. Not only had they landed on the boxes, but a couple of the earpieces had fallen through holes and into those two boxes.

Blaine reached around in the shadows to see if he could find whatever it was that had just fallen into his box. Maybe it was just a small piece of garbage, but it hadn't looked like it. He thought it might have looked like an earpiece. Before he could find out, Blaine felt his box drop off the end of the chute.

Tracey felt like she was in a blender as her box tumbled end over end, off the chute, down onto the dirt, and then into a hole in the ground. She had managed to grab the small item that had fallen into her box. It seemed to be an earpiece of some kind, but having a random earphone didn't much matter if she was about to be buried alive in a box.

Burly's smile was gone for only a second before it returned. Yes, Kimlee Broadstine had just knocked the box of earpieces out of his hand, but that would not stop him from giving his presentation. He had been trying very hard for a good long while to get some backing from some serious investors for his SmartDump idea, so he couldn't let this opportunity slip away, even if the investor that had shown up was a cranky-pants. Burly would continue his presentation, and he would do it with kindness and joy.

Sure, the front-end loader was still belching out noise and

Kimlee had rejected the offer to wear earphones. Even so, Burly continued to shout where Miss Broadstine could hear him, "Okay! So now you can see that all the trash has been pushed out of the truck, down the chute, and into the hole. That hole is what I call the 'esophageal hole.' And if you will follow me, I will show you why."

The trash man led the investor's daughter to a nearby set of stairs that led down to a door. Burly skipped down the stairs and he opened the door for Kimlee. After she hesitantly walked through, he followed her in to show her the next phase of his proposal.

"As you can see, Ms. Broadstine," Burly said, now away from the noise and talking at a normal volume, "I have set up this part of the SmartDump presentation to be viewed from a safe place."

In this underground room, Burly had put in a glass wall where it would be possible to clearly see the journey of the trash as it worked its way through the next few stages of SmartDump. "I call that hole in the ground outside the esophageal hole," Burly reminded Kimlee, "because it drops all the trash right here onto this conveyor belt, which leads to the first chamber. The first chamber works very much like the human stomach."

What was that faint buzzing sound? Blaine's box had gone dark again and it was churning around in circles, but what he had on his mind was this buzzing noise he'd begun hearing. The sound wasn't stationary. It was a buzz that seemed to be bouncing around on the inside of the box along with him. The buzz was on his right and then his left. Then, there it was dancing along his cheek. He reached up and swatted his cheek like he was trying to smash a mosquito. He got it! The buzz was now in his hand, and the buzz was . . . an earpiece!

After she'd heard it start crackling with sound, she had stuck it in her ear, which was when she heard Burly Scav's voice as clear as day. She didn't know how that was possible, but so far, nothing on this leg was making very much sense.

"The stomach is a 'J' shaped bag that churns and crushes food. It releases acid and gastric juices that begin to break down proteins. The stomach is lined with a thick mucus barrier so that these juices don't harm it. It stores food for several hours so that it can be slowly released into the small intestines. In the stomach, food is crushed and broken into a paste called chyme. It is this paste that enters the small intestines. And can you believe that the stomach can expand to fifty times its empty size to accommodate a meal? It's capable of holding up to four quarts!"

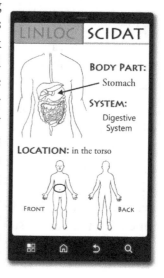

LINLOC   SCIDAT

BODY PART:
Stomach

SYSTEM:
Digestive System

LOCATION: in the torso

FRONT   BACK

"Oh, wow! That is so amazing!" Kimlee quipped in sarcastic enthusiasm. "I am so excited that I left my air-conditioned office at the crack of dawn this morning so that I could come out to this putrid landfill and learn about stomachs from a dirty trash man."

Burly was well aware that Kimlee was mocking him, but he went on with a smile, nonetheless. "Well, if you enjoyed that so much, Miss Broadstine, let me tell you a little bit about the small and large intestines, too. After the trash is churned and broken down for a while here in the Smart Dump stomach, it will move through a series of tubes and tunnels that can be compared to the intestines."

Kimlee exhaled in frustration.

"There are three sections in the small intestines, the duodenum, the jejunum, and the ileum. The duodenum is where chyme from the stomach is received, as well as pancreatic juices and bile. The jejunum is the section where digestive enzymes are produced. The ileum is covered with villi, which are little finger-like looking projections that absorb nutrients. So, all in all, food is largely digested in the small intestine with the digestive enzymes breaking down the food."

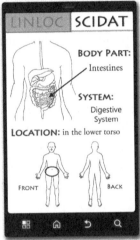

"Did you know that a microscopic view of the small intestine looks like a forest of villi? Nutrients from crushed-up food pass over these villi and are absorbed into the bloodstream. Anything left at the end of the small intestine will then pass through the large intestine. The main job of the large intestine is to absorb water, but it is also covered with millions of bacteria that break down part of the remaining waste and produce vitamin K for the body."

Burly paused for but a second and then continued, "After the food has been digested and all nutrients have been absorbed, the left over undigested waste, which is called feces, is released from the body at the anus, which is the end of the large intestines. Miss Broadstine, I told you all this about the stomach and intestines so you could better understand how Smart Dump works."

"As you can see right now," Burly said, while pointing through the glass wall, "the garbage is getting ready to tumble around in the stomach chamber. Through the use of some tested acid and mechanical crushing muscles, all this garbage will leave the stomach chamber after an hour or so as chyme-like paste. Then, it will be filtered through the two intestine phases and what comes out at the end will either be liquid or extremely compacted waste."

Blaine had just clearly heard every single word that Burly Scav had said. Under regular circumstances, that would be a good thing. But this was not a regular circumstance. He was stuck in a box underground, and he was about to be ground into paste by mechanical muscles and acid. As his beat-up box continued to move towards the chamber, he checked his phone one more time. Still no signal.

Separated from Blaine again, stuck in a box, no way to call Uncle Cecil, and now slowly approaching being turned into chyme: This is not how Tracey had imagined her summer going. Not before she had known about the zip lines and not after either. She felt herself surging forward, getting closer and closer to the trash stomach. It wouldn't be long now.

CHAPTER 11: THE END OF THE SMART DUMP

Kidney Punch!

Here they came, speeding up the dirt road in their shiny new pick-up truck. Smirk and Chili, the two meanest, nastiest... and coolest kids at school. Trevor didn't like them all that much, but if you wanted to be somebody at his high school, you had to be in with Smirk and Chili. His friendship with them was a fairly new thing. They weren't that great at being friends; they just seemed to pick on him all the time.

"Oh, well," Trevor thought. "That's the price I will have to pay to be popular."

When Trevor told them that he had to go to the landfill a lot for work, Smirk and Chili had acted like they thought that was a cool thing. They'd laughed and told him they wanted to come check it out. Trevor wasn't sure why, but regardless, here they came, driving a lot faster than seemed safe. Trevor was leaning up against the hood of his dad's trash truck, listening to music on his headphones.

Smirk, who was driving the pick-up, drove straight toward Trevor and the trash truck at full speed. At the last second he slammed on the brakes, sending the pick-up into a skid that finally stopped, just a few feet from Trevor.

Trevor smiled nervously and tried to act like what they had just done hadn't scared him, though his heart was beating a lot faster than normal. Smirk and Chili both jumped out of the truck with ornery looks on their faces. They were both big guys, a lot bigger than Trevor.

"What's up, Scab?" Smirk asked, as he walked up to Trevor. The boys had always called Trevor 'Scab' instead of Scav, his real last name.

Trevor just bobbed his head once in response, trying to be cool.

Chili ran up and shouted, "Kidney punch!" Then he proceeded to hit Trevor in the side just below the back section of his rib cage.

The Scav boy grunted and smiled, trying desperately to make sure that no tears formed in his eyes.

"That's Chili's new way of saying hello." Smirk laughed.

"And here is how I say hello," Smirk snorted and then spit on Trevor's shoe.

Trevor just put his hands in the pocket on the front of his hoodie, not really sure how to respond.

"So, this is the dump, huh?" Smirk turned from Trevor and scanned his eyes across the flat, dusty landfill.

"Yep," Trevor nodded. "This is it. Why did you guys want to come out here, anyhow? There's nothing fun to do."

"Nothing fun to do?" Chili said, sounding like Trevor's statement was preposterous. "Of course there is! There is always something fun to do!"

"Yep, Chili, there sure is," Smirk agreed. "And I've already spotted what we're going to do for fun here today!"

Smirk ran over to a nearby excavator and immediately climbed up into the cab of the dirty yellow and black machine. He ducked his head down and within seconds he reappeared with something shiny in his hand.

"Good news, boys!" he shouted. "We've got keys!"

"Oh, yes!" Chili screeched.

"Oh, no, Smirk and Chili want to take the excavator out for a joy ride." Trevor thought. He knew it was wrong, but the Scav boy followed along anyway. At that moment, impressing the two cool bullies was much more important to him than doing what was right.

Smirk ignited the growling engine, and the three teenage boys rolled off in the excavator.

"Man, there's a lot of levers in this thing," Chili noticed, and without asking how or what, he reached up and pulled one.

With a loud clank, the excavator's bucket fell down and smacked the dirt.

"Ha, ha, ha! Whoa!" Smirk roared, laughing enthusiastically. "That was gnarly, Chili! Pull something else! Let's see what we can do with this thing!"

Smirk slowed the vehicle down and Chili started pulling and pushing at all the buttons and levers on the dash. After testing them all out, he had, for the most part, figured out how to work the clawed bucket at the front of the excavator. All three boys laughed as Chili began digging a hole in the ground. Smirk and Chili's laughs were arrogant, but Trevor's laugh was nervous and fake.

"And our two last stages of Smart Dump are a filter and a holding tank that work together a little like the kidney and the bladder in the human body's urinary system," said Burly, still smiling and full of energy, even though Kimlee Broadstine had shown him nothing but disdain all morning.

He started to continue, but he stopped because he heard a sudden scraping noise.

"Do you hear that?" he asked Kimlee.

"Hear what?" the young investment broker snapped. "This whole place is making strange noises."

Suddenly, daylight appeared in the dirt over Burly's Smart Dump demonstration, on the other side of the glass wall from where he and Kimlee were standing.

"That's not good," Burly said, quietly.

Then in a terrible, loud, and sharp pitch, the glass barrier wall shattered into a thousand pieces. Dirt, glass, and garbage rolled down toward the two like a wave. In a flash, Burly grabbed Kimlee's arm and pulled her back through the doorway and up into the stairwell. The underground demonstration room quickly filled up behind them. He hurriedly helped her up the stairs and out of harm's way.

"Are you okay, Miss Broadstine?" Burly asked, concerned and in a cloud of dust.

"Get away from me!" Kimlee shouted ungratefully as she jerked away from the trash man. "This place is a nightmare! You are a nightmare!"

As Burly looked away from Kimlee, he saw what had made the scraping noise that had led to the shattered glass wall and what was, at this very moment destroying his Smart Dump site. It was an excavator, and it was being driven and operated by...Trevor?

How could this be? His own son, the son that he loved, was destroying all that he had worked so hard for.

Burly and Trevor's eyes met for one intense second, but then Trevor looked away. It was then that Burly saw there were two other boys with his son in the cab of the excavator. Even though all three had seen him, they weren't stopping their digging. They just continued to gouge a bigger and bigger hole into Burly's Smart Dump site.

The first smile of the day formed on Kimlee's face as she started to chuckle. "Well, Mr. Bernie, this turn of events is going to make it a whole lot easier for the Broadstine Group to make a decision on whether or not to invest in your Smart Dump project."

She wiped the dust off of her skirt suit and walked back over to her limousine, where the chauffer was already waiting with her door open. She put a foot in the car and then looked back at Burly and joked, "We'll be in touch."

Kimlee laughed as she disappeared behind a closed door and tinted glass. The chauffer gave Burly a sideways look before he got back into the driver's seat and drove the limousine out of sight. Burly watched as his chances of getting an investor drove away. Then, he turned back and stared at the excavator that was still destroying his dream, chunk by huge dirty chunk.

After the three boys had dug a crater sizeable enough for the moon, they turned off the machine and climbed out of the cab to greet the trash man heading their way. Trevor looked somber, but the other two teenage boys were laughing their heads off.

"Is that your dad, Scab?" one of the boys asked Burly's son.

Trevor nodded, and the two boys burst out in laughter.

The two big teenage boys walked up to Burly, still giggling. The first one spit on Burly's shoe and said arrogantly, "I'm Smirk."

Then, the next one shouted, "Kidney punch!" and punched Burly in the side.

"You actually just missed my kidney a bit and punched me in the ribs instead," the big trash man informed. "Kidneys are located on either side of the backbone, behind the stomach. You had the general area right, but you missed a bit and hit me in the ribs. Do you want to try it again?"

Chili had a confused look on his face. This had never happened before after he had dealt a kidney punch.

"Kidneys are key players in the urinary system," Burly launched into the science, unfazed by the disrespect from the boys.

"They are bean shaped organs. We normally have two of them. However, the body can survive with only one. Each kidney has three parts: the cortex, medulla, and pelvis. The kidneys' jobs are to filter waste products from the blood and to produce urine. You see, when the body digests food or when muscles burn fuel, some waste products are produced. The body uses what it needs and then has to get rid of the rest through the blood. The blood enters the kidney in the middle and spreads out into the medulla and cortex, where millions of tiny filtering units called nephrons filter the blood to remove waste products and excess water."

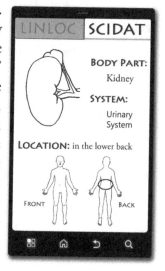

Smirk and Chili had stopped laughing and now had blank stares on their faces. Evidently, speaking scientifically to them was like speaking Latin to dogs. After a moment, they broke out of their trances and jumped right back into being who they were.

"Oh my goodness!" Smirk teased, looking back at Trevor.

"Your dad is such a nerd!"

"Yeah, Scab," Chili added. "Your old man is a loser! He's nothing but a science geek trash man!"

The two bullies walked back over to where Trevor was still standing.

"That excavator thing was a lot of fun," Smirk said to the Scav teenager. "What else can we do here? Is there something else we can drive?"

Trevor just shrugged.

"What about this?" Chili ran over to Burly's trash truck. "It looks like a piece of junk, but I bet it's a lot faster than that digger thing!"

This time Chili was the one who jumped in the cab first and fired up the vehicle's engine.

"Hey!" Burly shouted. "That's my trash truck! You already destroyed my Smart Dump. Don't steal my trash truck too!

"Oh, don't worry, Mr. Scab," Chili yelled from the driver's side window. "We're not stealing it. We're just borrowing it."

Chili laughed and then put the truck into gear. He steered around the shiny new truck that he and Smirk had driven in and then whipped the vehicle around close to Smirk and Trevor. Smirk quickly jumped up into the cab through the passenger side door. He looked down at Trevor, who was still just standing there in his hoodie.

"You coming, Scab?" he asked. "Or are you going to stay here with your loser dad?"

Trevor looked over in his father's general direction, but he couldn't bring himself to actually make eye contact with him. He looked back at Smirk and said flatly, "I'm coming with you, man."

Trevor hopped up into the cab with the two bullies, and then off the three teenage boys went in a cloud of landfill dust.

Burly Scav now found himself standing all alone. And for the first time in a long time, a frown appeared on his face. Not a mad frown, but a hurt and betrayed frown. The big man slumped to a seat right there in the dirt, unable to move from the shock of Trevor's choice.

All was quiet now. The Texas winds had stopped blowing momentarily across the flat Lubbock landscape, and Burly's Smart Dump project sat silent with all or most of its parts destroyed. Not a sound could be heard…except…wait, what was that?

It sounded to Burly like something was starting to rustle around in the dirt pile that the boys had created when they dug the big crater. Burly sat up straight, listening more intently, trying to figure out where the sound was coming from. Then, he saw it, on the other side of the hole in the ground from where he was sitting. There was some sort of old beaten up box and something was breaking it open from the inside.

"Well that's a strange sight," thought Burly to himself.

Then he saw that it wasn't a something, but someone. An arm came out of the box and then a head. It was a little girl! She looked to be about twelve or thirteen years old and a little scared, but very relieved. Burly jumped up and began running around the large ditch to get to the girl and help her, but before he could reach her, a hand popped out of the dirt right in front of him. Burly skidded to a stop.

The trash man dove down onto the dirt and started frantically digging. Within seconds, his digging revealed that the hand was connected to a person, and this person was a boy who was trying to claw his way out of another box, a box that was buried just under the surface of the dirt pile. Burly reached down with his big strong hands and pulled away a section of the box, creating a hole big enough for the boy to escape from. He then helped pull the boy up out of the box and the dirt and onto the ground. The boy, who looked to be about the same age as the girl, seemed to be a little shaken but fine, so Burly left him there to go and help the girl finish breaking out of her box. She was free except for her legs, which seemed to be stuck. Burly took one mighty swing and punched the box with a sledgehammer-like fist. The box shattered into an abundance of fractured pieces, and the girl was free. She stood up for a second, but then plopped down to an exhausted seat on the ground. Burly saw her make eye contact with the boy. Weak smiles of recognition formed on both children's faces, but neither said a word. They both seemed to be too weary.

Burly just sat there on the dirt pile at the edge of the big hole with the two children, in silence. He would give them as much time as they needed to recover from whatever it was that they had just gone through. He would let them speak first, if they even wanted to speak at all.

Bladders and Sludge Bogs

Tracey couldn't believe it. She was alive. She had survived the harrowing experience of being trapped in a box. A box that had been crushed, tumbled around, and then buried in an underground trash stomach. She was still shaking a bit from the experience, but she was now free and she was happy. To make her current situation even happier, Blaine had survived too! Evidently, he had been trapped in a box just like she had been, and he had gone through all the same scary things. Just like Tracey, he was very dirty and still shaking a little too. She could tell by the look on his face that he was relieved.

Tracey had been able to piece together a good idea of what had happened to her, Blaine, and to the trash man throughout the morning as she had been listening from inside the box. The only problem was: she knew what had happened, but not how or why? Their situation had improved and she was glad for that, but as Tracey looked at poor Burly, she realized his situation had gotten worse. At that moment, he looked pretty sad. "I'm so sorry Mr. Scav," she empathized.

"For what, dear?" Burly asked in curiosity.

"I am sorry that your Smart Dump got destroyed," Tracey responded.

"It sure did" Burly agreed, looking at the crater in front of them where the Smart Dump used to be just under the surface. "But how did you know that?"

The girl reached up and pulled an earpiece out of her ear, as did the boy.

"Wait a second!" Burly thought to himself. "I recognize those earpieces. Those are…but how?" He then reached up and felt the small clip-on microphone that was still on his shirt.

"We could hear what you were saying from inside of those boxes that we were stuck in," the boy spoke up.

"You could hear me?" Burly questioned.

A smile found its way to Burly's face again, but it was quickly gone again when he began to realize what the children had just been through. He and Trevor must have somehow unknowingly loaded them into the back of the trash truck on their morning run. The boxes that the kids had been in had somehow gone through and survived the mechanical teeth in the truck,the esophagus chute, and the underground stomach.

"Oh, you two!" Burly said, compassionately alarmed. "I am so sorry that you got thrown in the garbage!" The trash man looked at the two dirt-covered children. "Who cares about the Smart Dump! The important thing is that you two are okay! I was sad that my son and his friends tore up my site with the excavator, but in doing so, they unwittingly dug you two out in the process. So it's okay! Not only is it okay, it's miraculous!"

All three paused for a moment and dwelt on the morning's sequence of events.

"What are your names?" Burly asked the children. "And how did you get stuck in the boxes?"

"We are the Sassafras twins," the boy answered, "Blaine and Tracey Sassafras."

"But we really don't know how we got stuck in the boxes," Tracey added. "One second we were out in the open, and the next second we were inside the boxes."

"Well, regardless of how you got stuck inside the boxes" Burly said. "I am just so sorry that I didn't see the boxes and pull them out before they went through my Smart Dump."

"It's not your fault," Tracey assured the trash man.

"Yes," Blaine agreed. "It's not your fault. And besides, I really think your Smart Dump is a great idea. I would like to know more about it."

"It looks like you're the only one." Burly sighed. "Kimlee Broadstine from the Broadstine Investment Group didn't seem too impressed, and my son Trevor, whom I've told all about it, doesn't seem too impressed either. Plus, he thinks his old man is a joke."

"You're not a joke." Tracey tried to encourage the man. "I agree with Blaine. the Smart Dump is a great idea! Is there anything else that you didn't tell Kimlee about it that you would like to tell us?"

"Yeah, we would like to know all about it!" Blaine cheered. "The last thing I heard in the earpiece, before being dug out, was about that filter that kind of acted like a kidney."

"Wow!" Burly smiled. "You two really were listening!"

The twins nodded. Burly looked down at the big hole again that the teenage boys had dug.

"I don't know if anything is left," he exhaled. "But we can take a careful climb down and look."

The three slowly made their way down into the crater, carefully stepping over sliding dirt, garbage, and broken mechanical pieces. They found the remnants of the underground viewing room, but the room was now so full of dirt there was only enough space between the ceiling and the rubble to crawl. Burly crawled in first, but then paused and looked back at the twins.

"Are you two sure you want to crawl into this tiny space after your box experiences?"

Both Blaine and Tracey bravely nodded that they did, so Burly led on. After several yards of crawling on their hands and knees, the three found that the room opened up a bit, and they were able to stand. Burly pulled a small flashlight out of the utility pocket of his jeans. He clicked it on and shown it around to see if there was anything left of his Smart Dump.

"Well, Blaine and Tracey," he declared after a while, "it looks like everything is destroyed except for possibly the holding tank."

"You mean the holding tank that acts kind of like the bladder in the human body?" Blaine asked, trying to confirm what he had heard Burly say earlier in the earpiece.

"Yes, Blaine," the trash man affirmed. "That's exactly right. This tank was designed to catch any liquid that was squeezed and compressed out of the garbage. Then it slowly releases the liquid into a water treatment machine. It's not exactly like the human body of course, but it's pretty close. The bladder in the body is a holding tank for urine. It is just one part of the urinary system. The other parts that make the urinary system are the kidneys, urethra, and ureters. The job of the urinary system is to remove waste and extra water from the body, and to produce urine.

"So how do all those parts of the urinary system work together?" Tracey asked.

"Wow!" Burly exclaimed happily. "It's so fun that you guys are so interested in all of this science!"

Blaine and Tracey smiled. It still surprised them too that they were so interested in science when, only a couple of weeks ago, they had loathed it so.

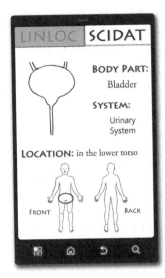

"To answer your question, Tracey," Burly began, "urine that is collected in the pelvis in the kidney passes through the ureters. The ureters are eight to ten inches long. Small amounts of urine pass through every ten to fifteen seconds. The right and left ureters bring urine from the right and left kidneys and dump it into the top of the bladder. Urine is composed of water, salt, ammonia, waste products and pigments. The more water you drink, the more water is contained in your urine and the lighter colored it will be.

The adult bladder is the size of a small fist, but it can expand to hold about two cups. When it reaches half capacity, the bladder will send a message to the brain so that you feel the urge to go to the bathroom. A ring of muscle called the sphincter holds the opening to the urethra closed until it receives a message from the brain to relax. Then, with the aid of gravity, urine is released into the urethra and carried outside of the body. The walls of the bladder can also contract to squeeze urine out. Pretty interesting stuff, huh?"

It was a little bit gross, thought the twins, but it definitely was interesting.

Burly sighed as he looked around a little more with his flashlight. Then, the three crawled out of the underground room and back up onto the dirt pile at the rim of the ditch. Burly looked like he was about to say something else to the twins, but then his face froze as he stared off into the distance behind the twins. The Sassafrases turned to see what he was looking at.

"It's Trevor," Blaine and Trace heard the trash man say. "He's coming back." Sure enough, here came the teenage boy, walking back from wherever it was that he had gone. He was by himself and he was walking rather slowly, looking down at the ground, with his hands in his front pocket and his hood covering his head. The three sat in silence, waiting for the boy to reach them. When Trevor finally did reach the dirt pile, he just stood there silently in front of them for a few seconds, shuffling his feet and fidgeting with a pair of earphones.

Then, all at once, he looked up and made eye contact with his father. The twins could see that Trevor had sincere regret in his eyes. The teenager looked like he was about to say something, but was interrupted by the sound of an approaching vehicle. The twins immediately recognized the vehicle as a trash truck. The truck sped up to the ditch faster than one would think that a trash truck could go. Then upon reaching the ditch, the driver slammed on the brakes, and the truck came to a quick stop. Two big teenage boys opened up the doors and hopped out onto the ground. The twins assumed that these must be the two bullies they had heard Burly talking to when they were listening from inside the boxes.

"I thought you were going to stick with us Scab." One of the big boys asked Trevor, "Did you change your mind and want to run back to your daddy?"

"And who are these two kids?" the other asked, pointing at Blaine and Tracey. "Is this your baby brother and sister? What is this, some kind of Scab family reunion? Hey Smirk," the second boy suggested to the first, "let's introduce ourselves to the two little Scabs."

He started running toward Blaine and Tracey and shouted "Kidney punch!" but before he could reach the twelve-year-olds, Burly stood up and stepped in his path.

"Oh no you don't, Chili!" the big pot-bellied trash man warned.

The boy named Chili lowered his shoulders and tried to run through Burly to get to the twins, but when he hit the trash man it was like he hit a brick wall. Upon impact between the two, Chili fell to the ground, but Burly didn't move at all. Chili quickly hopped back up and tried to act like his smash in with Burly hadn't hurt at all.

"Okay, Mr. Scab," he grunted, a little shaky. "We'll just introduce ourselves and say hello to those kids later." He wobbled back over to where the other bully was standing next to Burley's son.

"So, I ask you again, Scab." Smirk growled intimidatingly at Trevor. "What's it going to be? Are you going to hang with us or are you going to stay here with your loser dad?"

Trevor looked down at the ground again, as if trying to make a decision, but after only a second he looked up and answered, "I'm staying here with my dad."

Both Smirk and Chili's faces showed that they thought Trevor's decision was ludicrous.

"What!" Smirk exclaimed. "You're going to choose your dad over us?"

Trevor nodded that he was.

"Just look at him, Scab." Smirk pointed at Burly. "He's a loser. He drives a trash truck and spends most of his time at a landfill. And did you hear all of that nerdy science stuff he was blabbering about earlier? He's just a big fat loser geek!"

The twins couldn't believe how hateful these two bullies were being. On the other hand, Burly looked completely unfazed by the disrespectful words of the teenager. He was simply looking intently at his son to see how Trevor would respond. A sudden resolve filled the young Scav's face as he looked right at Smirk.

"My dad is not a loser," he said boldly to the bigger boy.

"Yes, he is!" the bully laughed arrogantly.

"No he's not," Trevor countered. "My dad is the greatest man that I have ever known."

Burly's mouth dropped open in surprise.

"Seriously?" Smirk asked. "That's what you really think?'

"Yes" Trevor answered. "That's what I really think. I not only think it, but I know it. My father, Burly Scav, is a great man. Life has never given my dad any special advantages, but he has taken what he has been given and he has made the most of it. He's smart, wise, and strong. Not only is he strong enough to clobber the two of you into the ground right now if he wanted to, but he is strong in character. He treats people with kindness and respect, even when no one is looking. My dad is a great man. The two of us don't always know how to communicate with each other the best, but that man is my hero and I want to be just like him."

Burly could not believe it. He had never heard Trevor say anything like what he had just said. He knew that his son loved him deep down, but he had just assumed that Trevor thought he was a failure. He had no idea that his teenage boy thought that he was a great man.

Evidently, Smirk and Chili couldn't believe it either. Their faces showed just as much surprise as Burly's. They were not used to being rejected like this.

"Well, I have news for you Scab!" Smirk shouted right into Trevor's face. "You already are just like your dad. You are a loser nerd, just like he is."

"Yea!" asserted Chili, as he stepped up and put a finger right in Trevor's chest. "You are a loser, Scab. You just wait until school starts again after the summer. You are going to get it, buddy."

"Yes you are," Smirk agreed. "We were really thinking about bringing you into the in crowd with us, but after this, you are going to find out what it is really like to be on the outside."

Together, Smirk and Chili shoved Trevor to the ground, and then they turned to get into their truck. Smirk fired up the powerful sounding engine of the shiny new truck, pushed on the gas, and off the two bullies went.

Trevor looked over at his father. "I'm sorry dad," he repented quietly. "I'm sorry that I didn't stand up to those guys sooner. I'm sorry that I let them destroy your Smart Dump site. Now everything that you worked so hard for is gone, and it's all my fault." Trevor heaved a big sigh and looked like he was holding back tears. "I'm not even worthy to be called your son. I'm sorry, dad."

Trevor's last statement barely made it out of his mouth before Burly stepped over, picked his son up off the ground, and wrapped him up in a huge embrace.

"I love you, Trevor," the big unshaven trash man said. "I love you no matter what, and you are always worthy to be called my son."

The two Scavs stood in an embrace, patting each other's backs and sending up small clouds of dust. It was a wonderful father/son moment that would have lasted longer if a loud crashing sound hadn't rang out into the air.

"Well, that's not good," Burly commented, when he saw what had made the noise. "It looks like ole' Smirk and Chili have just crashed that shiny new truck of theirs into the sludge bog."

The twins looked in the direction of the noises' source to see what Burly was talking about. A couple hundred yards or so from where they were, the bed and back two wheels of a truck were sticking up into the air.

Burly started to chuckle and looked at Trevor. "Those two are a mess aren't they?"

Trevor nodded, laughing as well.

"I guess we should go pull them out," Burly suggested. The trash man, his son, and the Sassafras twins all hopped in the cab of the trash truck and headed over to the stuck truck. On the way over, Burly explained that the sludge bog was a small pool where he had been collecting trash sludge, so that he could test out different methods of water purification.

When they reached the immobilized truck, the four could hear Smirk and Chili yelping out like helpless puppies.

"Help! Get us out of here! What is this stuff? It stinks! Oh man, it stinks!"

Their truck was almost halfway submerged in the greenish brown sludgy liquid, and it was slowly sinking deeper.

"Hold on, boys," Burly shouted, smiling. "Let me hook you up to the winch. You'll be out of there right quick."

"Please hurry! Oh hurry! Hurry! Hurry!" the two big teenage boys screamed.

In a matter of minutes, Burly had used the winch to pull Smirk, Chili, and their pickup truck out of the bog. The two bullies now stood dripping in sludge, facing the four.

"Your cousin is going to kill us!" Chili screeched to Smirk. "We just totally crashed his new truck."

Smirk looked like he was about to cry.

"You mean that wasn't even your truck?" asked Trevor.

Smirk and Chili shook their heads 'no.'

Burly walked over and opened the slimy truck's hood.

"Looks like the engine's shot," he confirmed. "I think we are going to have to call a tow truck to get this thing out of here."

Burly pulled out a cell phone to make a call, but before he could dial, the phone started ringing.

"Yello," the trash man answered cheerfully, as he walked around to the other side of his truck to talk to whomever it was that had called.

Smirk and Chili looked humbly at Trevor.

"Thanks for pulling us out, man. I guess you and your dad aren't losers after all," Smirk acknowledged, as Chili nodded in agreement.

"No problem." the young Scav shrugged.

After a couple of minutes, the five kids heard a shout of joy, as Burly came running back around the trash truck.

"Trevor!" Burly exclaimed. "You're not going to believe it!"

"What, dad, what?" Trevor asked, as his dad grabbed his shoulders and started shaking him in joy.

"You are never going to guess who that was!"

"Well then, tell me!"

"That was Broderick Broadstine!" Burley said in elation. "The father of that lady that was here earlier, checking out my project! He is the president and founder of the Broadstine Investment Group. I thought I had no chance Trev, no chance at all of getting an investment after what happened here earlier, and how Kimlee treated me, but…"

Burly let go of Trevor's shoulders and started dancing in a circle.

"But what, dad, what?" the son questioned.

"We did it, son! We got it! Mr. Broadstine is going to invest in Smart Dump!"

"He is?" Trevor asked.

"Yes he is!" Burly answered joyfully. "He even apologized for Kimlee's behavior, and said that she's a good kid and he loves her, but she is just going through a rough stage as a young professional. He said that she came stomping in the office, complaining about her morning here at the landfill in detail. He loved all that he heard her say about Smart Dump, so much, that he's going to invest!"

"That's great, dad!" Trevor exclaimed.

The Scavs gave each other another big slapping hug, while smiling from ear to ear. Blaine and Tracey smiled too. They were happy for their local expert and his son, and they were happy for themselves. In what had been one of the scariest mornings of their lives, all had worked out for the best for everybody, except maybe Smirk and Chili. Burly had gotten his investment. Trevor had gained the respect of the bullies. The Sassafras' had survived being trapped and buried in those strange boxes, and to think that all this had happened before lunch!

They were happy that it had all worked out okay, but they were still troubled about what had happened to them. Before they used their phones to zip to the next location, they were going to use them to call a certain red-headed scientist that they knew.

CHAPTER 12: INTERNING IN BANGKOK

Reproduction Realized

"Well, wee willy willwackers, President Lincoln," the eccentric scientist laughed, saying to his pet prairie dog, "I had forgotten how hard it is to fit all of this inside of Socrates and Aristotle!"

Cecil Sassafras was fumbling around with yards and yards of spongy and bouncy fake intestines, trying to stuff them correctly into the frames of his two plastic skeletons. President Lincoln was up on Cecil's desk, using the computer's mouse to scroll through all the SCIDAT data and pictures Blaine and Tracey had just sent in.

"Aren't those twins doing a wonderiffic job, Linc?" Cecil declared proudly to his furry friend. "Location after location and subject after subject, they continue to deliver!"

Suddenly, the pocket of the scientist's white lab coat started to vibrate.

"Sweet sizzling squanto sticks!" Cecil said, somewhat startled. "Somebody is calling me." He pulled the smartphone out of his pocket and pushed the answer tab.

"Howdy hooty!" he greeted cheerfully.

It was Tracey.

"Did you get all the data and pictures from the digestive and urinary systems?" the twelve-year-old girl asked.

"Sure did, Trace-a-fras," Cecil confirmed. "You and Blaine are doing great! Are you two ready to go to the next new and exciting location?"

Tracey paused and then answered half-heartedly, "Maybe."

"Maybe? Maybe?" Cecil responded, puzzled. "Why just

maybe?"

"We've had a lot of crazy things happen to us since we started this zip lining scientific journey, Uncle Cecil," Tracey unloaded. "A lot of scary things, but none of our previous scary situations compare to what we went through this morning."

"What happened?" Cecil questioned.

"We got buried alive!" Tracey exclaimed.

"Buried alive?" Cecil repeated, nearly jumping out of his bunny slippers. "How did that happen?"

"We don't really know," Tracey answered. "We were on the Great Wall of China, and we had just sent you the data on the muscular system. LINLOC had already given us our next location and we were cinched up in our harnesses, with our calibrated carabiners hooked to the lines and ready to go. Then, suddenly, Blaine and I were trapped inside two boxes. It was so scary, Uncle Cecil. It was cramped and dark, and no matter how hard we tried, we couldn't break out of them. We couldn't call you, either. Our phones had no reception inside the boxes. Our local expert and his son accidentally buried us in the boxes, but in the end, they were the ones that rescued us as well."

Cecil was taken aback by Tracey's story. He didn't mind if his niece and nephew faced a little peril in the adventures of their summer, as challenging situations are good for a person's soul, but he didn't want them to be scared. He didn't want them to lose heart, and he didn't want them to go back to hating science.

A born encourager, Cecil Sassafras wanted to reassure Tracey. "Tracey, so far this summer, you and Blaine have zippity zapped all the expectations that President Lincoln and I had for you out of the water. You have excelled at learning science, and you have stuck true to your Sassafras blood by never giving up. Not only have you not given up, you have helped so many others along the way. I just know you can keep doing it!"

THE SASSAFRAS SCIENCE ADVENTURES

Tracey paused. The few words Uncle Cecil had spoken had lifted her spirits, but she still felt worried.

"But what about the boxes?" she asked. "What if we get trapped in those things again and there is no one there to rescue us? Uncle Cecil, do you think it was the Man With No Eyebrows? Do you think he's the one who put us in those?"

"Ahhh, the Man with No Eyebrows," Cecil thought to himself. "I do wonder who that man is and why he keeps showing up along the twins' journey."

He rubbed the whiskers on his chin before he responded to his niece, "Well, regardless of who that guy is and what he has up his sleeve, he is no match for the Sassafras twins, right?!"

Tracey paused and managed a small laugh. "Right, I guess."

"Okie dokie smokie pokie," Cecil said energetically. "Then go out there and get it done! Press on! You two can do it!"

"Thanks, Uncle Cecil," Tracey responded.

She pushed the button to end the call. She turned to look at Blaine, who was already in his harness.

"Onward?" he asked.

"Onward," Tracey confirmed.

As she opened up her backpack and pulled her harness out to put it on, Blaine read the information as he now found it on LINLOC.

"Our next location is Bangkok, Thailand: Longitude 100° 28' E, Latitude 13° 43' N!" he announced, excited. "Our local expert's name is Dr. Apple, and we will be gathering data on the reproductive and immune systems."

"Dr. Apple. Bangkok, Thailand," Tracey repeated. "Sounds interesting."

The twins now hung in the air with their harnesses on, their carabiners calibrated, and their fingers crossed that they wouldn't get trapped in any more boxes.

Whoosh! Off they went into the light at rip-roaring speed, free and soaring. No boxes, no darkness, only light; this is how it was supposed to work! The Sassafrases were relieved and delighted that things seemed to be happening normally again.

They landed with a jerk, slumped to a halt, and waited for strength and sight to return. When they regained their senses, they saw they'd landed in some kind of storage room. By the looks of it, it was a hospital storage room. There were wheelchairs and crutches, medical instruments of all kinds, and a rack of hanging white lab coats like doctors wear, not unlike the one Uncle Cecil was always sporting.

Blaine stood up, walked over to the rack, and pulled a lab

coat off of a hanger and tried it on for size.

"Look, Tracey," he showed off. "I'm a doctor. Instead of calling me Blaine, you can now refer to me as Dr. Sassafras."

Tracey stood up. "I'm not going to call you Dr. Sassafras," she asserted, rolling her eyes but smiling at the same time. "Besides, that lab coat is way too big and baggy for you. You look ridiculous."

Blaine grabbed another lab coat off the rack and tossed it to his sister. "Oh, c'mon, Trace, try it on. It may look ridiculous, but it's fun to feel like a doctor.

Tracey grabbed the lab coat and put it on.

"See?" Blaine smiled. "It feels nice, doesn't it?"

Blaine now started reaching around on one of the shelves and found a couple of surgeon's masks. He tossed one of those to Tracey as well. Both twins strapped the lightweight masks over their noses and mouths.

"We aren't here to play dress-up," Tracey admonished in a slightly muffled voice from behind the mask. "We're here to learn some science."

"True," Blaine agreed. "But who said we can't be dressed up like doctors while we learn science?

Tracey laughed at her brother's antics and watched as he stepped over and swung open the storage room's door out into the hallway. Blaine took one comically confident step out into the hallway, but the swinging door abruptly smacked into something, and then they heard someone yelp.

"Oh, no, Blaine," Tracey exclaimed. "I think you just hit someone with the door."

Tracey followed Blaine, who was already peeking around the door to see who or what he had just hit. To Blaine's shock and embarrassment, he had knocked over a man who had been walking with crutches and was all wrapped up in different kinds of bandages and several casts.

"I am so sorry, sir!" Blaine apologized from behind the surgeon's mask. "Let me help you up."

Blaine and Tracey both bent down and tried to help the man get up off the floor, but the man was heavy and his balance wasn't very good. So, all the Sassafrases managed to do was get him about halfway up and then accidentally knocked him back over as they fell on top of him.

"Oh, geez," Blaine blurted out. "I'm sorry, sir. Again, I'm so sorry."

"Well!" the Sassafrases heard a female voice say. "What on earth is happening here?"

They looked over to see a pretty woman with shoulder length brunette hair who looked like a doctor walking in their direction. "What are you two doing to this poor man?" she asked.

The twins quickly rolled off of the man, but remained speechless. As the woman got closer, Blaine and Tracey saw that the doctor's name was embroidered on her lab coat. "Dr. Apple," it clearly read. This was their next local expert. The doctor picked up the man's crutches and then successfully helped him back up to his feet.

"Oh, it's just you, Nart," Dr. Apple acknowledged the man. "Sorry our new interns couldn't get you up off of the ground. Are you okay?"

The man nodded and even managed a smile. He then turned and hobbled down the hallway on his crutches. Dr. Apple looked the Sassafrases up and down.

"You two are new medical interns, aren't you?" she asked.

"Well...uh...actually..."Blaine stammered.

"My goodness," Dr. Apple shook her head interrupting Blaine. "The interns they send us just keep getting younger and younger. Either that, or I'm just getting older."

The doctor sighed. "Well, follow me, you two."

"Oh, snap," both twins thought. "We made a terrible first impression on our local expert and now we're in trouble. We're probably going to get busted for getting dressed up like doctors."

Dr. Apple led them over to an elevator. She pushed the up button. Once inside the ascending elevator, she assured, "You guys don't need to worry about Nart. He'll be fine. He's basically a permanent fixture here at our hospital. He has the longest streak of bad luck going that I've ever seen. He seems to be perpetually getting into accidents. No worries, though, one of our neurologists recently concluded that Nart doesn't feel pain. We keep fixing his broken bones and stitching up his wounds, but he can't feel a single one of them. Plus, I think he actually likes being at the hospital."

The elevator doors opened on the fifth floor.

"This is the women's clinic that I run," Dr. Apple declared. "You two follow me and I'll show you the ropes."

"Maybe we aren't in trouble," the twins thought.

Dr. Apple grabbed a chart from an approaching nurse and then walked over to a curtain and pulled it open while reading the chart. The open curtain revealed a Thai lady with rosy cheeks and long hair who was sitting up in a hospital bed.

"Mrs. Kanthachai?" Dr. Apple asked the woman, confirming her name.

Mrs. Kanthachai nodded that, yes that was her name.

"An apple a day keeps the doctor away. That's what they say, anyway, but hey, an apple IS your doctor today!" Dr. Apple quoted this nifty little updated poem while pointing at her nametag. Mrs. Kanthachai smiled pleasantly.

"What seems to be the problem this evening?" the doctor asked kindly.

"My tummy feels very icky," the woman answered. "I think maybe it was something I ate."

Dr. Apple looked at the twins and spoke to them like they were indeed medical interns. "The nurses have already taken all of Mrs. Kanthachai's vitals. Now I will do just a couple of quick tests to see what is wrong with her stomach."

The twins nodded like they understood, but they really weren't sure if they were supposed to comment and assist, or if they should just stay quiet and observe. The latter seemed like the best option for now.

After a couple of minutes of poking and prodding, Dr. Apple stated, "Well, Mrs. Kanthachai, I don't think it's something you ate."

"You don't?" the Thai woman responded.

"Nope," Dr. Apple shook her head and smiled. "Mrs. Kanthachai, I believe that you're pregnant!"

Mrs. Kanthachai looked confused. "Pregnant? How did that happen? Both my husband and I have been eating well and exercising. How long will this take to cure?"

"About nine months," the doctor replied.

"Do you think my husband has it too?" Mrs. Kanthachai asked, alarmed.

"Highly unlikely," Dr. Apple said.

"Well, what can I do to treat it?" the poor confused Thai woman questioned.

Instead of answering immediately, Dr. Apple just looked at her patient and tried to read her face. Originally, she thought Mrs. Kanthachai was just joking, or at most was shocked by the news she'd just received and had momentarily forgotten how things worked in the world. But as Dr. Apple looked into the woman's eyes right now she realized Mrs. Kanthachai sincerely didn't know what was going on. Dr. Apple took a deep breath and began the explanation.

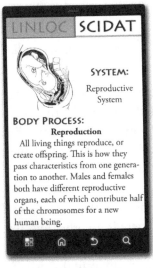

LINLOC SCIDAT

SYSTEM:
Reproductive System

BODY PROCESS:
Reproduction
All living things reproduce, or create offspring. This is how they pass characteristics from one generation to another. Males and females both have different reproductive organs, each of which contribute half of the chromosomes for a new human being.

"Mrs. Kanthachai, all living things reproduce, or create offspring. This is how they pass characteristics from one generation to another. Males and females both have different reproductive organs, each of which contributes half of the chromosomes for a new human being. The reproductive organs in males produce sperms, and in females they produce eggs. Both the sperm and the egg are single cells that contain one set of twenty-three chromosomes. These cells meet to form a complete genetic package, which we call a zygote. This zygote then grows and develops over nine months inside the mother's body in an organ called the uterus. The zygote starts out looking like a ball of tightly packed cells and ends up looking like a miniature version of an adult human. While the zygote is growing it is protected by the amniotic sac and is fed through the umbilical cord. Once this zygote, or tiny human, is delivered, it takes its first breath of air and begins supporting its life on its own." Dr. Apple paused and smiled again. "Mrs. Kanthachai, you are going to have a baby."

Mrs. Kanthachai looked shocked all over again, but this time the shock was accompanied by a look of understanding.

"So being pregnant and having a zygote inside of me means that I am going to have a baby?" the Thai woman confirmed.

Dr. Apple, still smiling, nodded her head yes. Mrs. Kanthachai started gently crying. Dr. Apple reached a hand over to comfort her.

"I'm crying because I'm happy," Mrs. Kanthachai beamed. "And my husband will be so happy too."

"I will be your doctor throughout your pregnancy," Dr. Apple informed. "That is, the nine months or so that the baby is inside of you. I will also assist you with the delivery of your little one."

"That sounds wonderful," Mrs. Kanthachai replied thankfully.

"Newborn babies are completely dependent on their mothers," Dr. Apple added. "They are fed only milk, they can't move much on their own, and they sleep most of the time. But as they grow, their muscles, digestive system, and the rest of their body grow and develop so that they are able to do more and more on their own. This whole process, from now through development, requires a lot from the mother, to say the least, but there is no effort that is rewarded as much as motherhood. There is no other love like the one that you will have for your baby. Myself, the nurses, and the rest of the staff will be here to help you in any way that we can throughout the entire process."

"Thank you!" Mrs. Kanthachai replied gratefully, still crying tears of happiness.

The Body's Blueprints

The Sassafras twins followed Dr. Apple around, quietly observing everything the respectable doctor did as she cared for, diagnosed, and helped several other women. Dr. Apple had just mentioned to the twins something about a coffee break when a tall, fit Thai man stepped off of the elevator hastily, like he was looking for someone. When he spotted Dr. Apple, a look of recognition came to his face, and he walked quietly towards her.

"Olivia!" the man whispered hurriedly, when he reached the doctor.

"So evidently this man knows Dr. Apple," thought the twins. "And seemingly Dr. Apple's first name is Olivia".

"Chanarong!" responded Dr. Apple. "Tell me you have something for me!"

The man nodded, but looked neither happy nor sad.

"Olivia, I have good news and bad news."

Dr. Apple twirled her hands in front of her, as if telling Chanarong to get on with it.

"The good news is: we have found her and she is alive," Chanarong shared.

"And the bad news?" Dr. Apple enquired.

"The bad news is: the man that has kidnapped her is none other than Kingman Nawarak."

Dr. Apple's face suddenly had a shocked and angry look on it. "Kingman Nawarak!" the doctor exclaimed. "But that's…he's… this is not good."

"No it's not," Chanarong stated gravely.

A look of sorrow covered Dr. Apple's face, but it was soon replaced with a firm resolve.

"Well, what are you and your department doing to get her back?" the doctor asked.

"Nothing," Chanarong added flatly.

"Nothing!" Dr. Apple retorted, as if Chanarong's answer was completely unacceptable. "How can you be doing nothing? You have to go rescue her and bring Kingman Nawarak to justice!"

"We can't," Chanarong acknowledged, with more than a hint of disgust in his voice. "Kingman Nawarak has money and power, and he has been successfully eluding arrest and imprisonment for years. Everyone in Bangkok knows what an evil man he is and how he makes all of his money, but neither the Thai government nor the Thai police force has been able to stop him or prove any wrongdoing on his part." Chanarong stopped and shook his head.

"Olivia, I am in the process of putting together a task force that will attempt to take Nawarak down, and rescue any and all that he has kidnapped. But as you well know, bureaucracy and red tape will prohibit this task force from being as quick as it should be. I am sorry to say that my hands are tied, Olivia."

Dr. Apple clenched her fists and gritted her teeth. "Well then, I will go after her" she resolved. "I will go after her, and I will bring Kingman Nawarak to justice.

A half smile formed on Chanarong's face. "I knew you were going to say that" he expressed. "We have been friends a long time now, and I know you well enough to know that nothing I can say will change your mind. So that is why…even though it may get me fired…I am going to help you."

"You're going to help me?" Dr Apple asked.

"Yes I am," Chanarong responded. He took off the bag that he had slung over his shoulder and handed it to Olivia. "Inside this bag you will find precise directions to Nawarak's compound outside of the city as well as a blueprint of that compound. Also, I've added a handful of supplies that could aid you in your rescue attempt. I know you like to help people, not hurt people, so you will have to be creative in your rescue approach.

Chanarong stopped for a second and took a deep breath. He looked right into his friend's eyes and, sincerely concerned, said, "Olivia, you know this is going to be next to impossible, right?"

Dr. Apple nodded, but not even the slightest bit of resolve left her face. Chanarong nodded, too, knowing his friend's heart and mind were set.

"Be careful," he cautioned. Then he turned and left, choosing to take the stairs this time.

Blaine and Tracey were still pretty much clueless as to what was happening. Dr. Apple motioned for them to follow her, and they heeded her instructions. She led them to a supply closet, much like the one they'd landed in. Then, she had them close the door.

She immediately opened up the bag her friend had given her and began rummaging through its contents. Once she found two rolled up pieces of paper, she pulled them out, slid off the rubber bands, unrolled them, and had the twins hold them open on the floor. The first was a map of Bangkok and its surrounding area. Just outside the city, a small area was circled in permanent marker. Dr. Apple put her finger on the marked spot.

"There you are, Kingman," she pointed, as if talking to the villain. "You should be easy enough to find.

Next, she looked over at the other large piece of paper, which was the blueprint of Nawarak's compound. After a quick glance she avowed, "Hang in there, Lawana. We're coming to get you."

"Who is Lawana?" Tracey asked.

"She's my friend," Dr. Apple grinned, but then sorrow quickly filled her face. "Now she is a captive of this character, Kingman Nawarack. Kingman. What kind of name is that, anyway? Word on the street is that he gave himself that name, but he is neither a king nor a man. He is a monster and the three of us are going to take him down!"

The Sassafras twins both pointed at themselves as if questioning that they were included in 'the three' Dr. Apple had just mentioned.

"Of course you two are going to help me!" Dr. Apple confirmed.

She patted the twins on the back and stood up. "What better way to start a medical internship than to go rescue a pregnant widow from the clutches of a madman?"

"Lawana is a widow?" Tracey asked compassionately.

"And she's pregnant?" Blaine added.

Dr. Apple nodded her head slowly and had a look in her

eyes like she was thinking back to events of the past.

"Lawana was my first Thai friend. I met her right here in Bangkok when I was with the Peace Corps. After two years of friendship, I went back to the States to finish med school, and Lawana married her childhood friend, Mongkut Planisong. Mongkut was a wonderful man. He was handsome and kind, and he truly loved Lawana. He didn't have a lot of money or anything material to offer her, but she was more than happy to marry him. Tragically, not even five years after they were married, Mongkut was badly injured in a construction accident. I was back in Thailand when the accident happened and was working here at the hospital. The doctors in the E.R. did all they could to save him, but Mongkut's injuries were too severe, and he died.

"Lawana was devastated, but her heartbreak was eased when she found out just days after the accident that she was pregnant. She was going to have Mongkut's child. I wept with her over the loss of her husband, and then I rejoiced with her over the news of her pregnancy. But now..." Dr. Apple paused and went from reflective to determined. "Now Kingman Nawarak has her. A man who kidnaps helpless women and children for his own gain. At this point, Lawana is nine months pregnant and due to give birth any day, and I'll be a monkey's auntie if I'm going to let her have that baby in Nawarak's compound. Lawana is my friend, and that baby of hers has good genes, Mongkut's genes. He or she will not be born Kingman's captive!"

"Babies can be born wearing jeans?" Blaine asked absentmindedly, not meaning to be silly.

Tracey ducked her head, embarrassed at her brother's ridiculous question, and Dr. Apple looked at Blaine like he was crazy.

"Are you serious?" the doctor asked the twelve-year-old. "What are they teaching you youngsters in med school these days? Not 'blue jeans', 'genes'."

THE SASSAFRAS SCIENCE ADVENTURES

Blaine stared at the woman blankly.

"Genes? Chromosomes? Codons? DNA? Any of this ringing a bell?"

Blaine remained frozen.

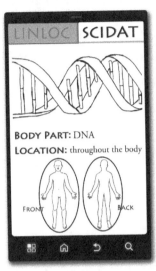

"OK, then, let me start from the beginning," the respectable doctor sighed. "First, we have DNA, otherwise known as deoxyribonucleic acid. But that's too hard to say, so we just say DNA. This stuff is so tiny that it can't be seen unless you use a very powerful microscope. If you could see it, you would see what looks like a twisted ladder. Scientists refer to this ladder as the double helix. The DNA ladder is composed of rungs that are made from two letters of the DNA alphabet, which consists of only four letters: A, T, G, and C.

"Each letter has a unique puzzle-like shape, so that A and T fit with each other to form a rung on the ladder, and G and C fit together to form a rung on the ladder. As you read the DNA ladder, the letters combine to form three-letter words called codons. These codons then combine to form sentences, which we call genes.

"Genes are the basis for your chromosomes, and chromosomes give your body a blueprint, or set of instructions for life. Every human has twenty-three pairs of these DNA chromosomes that tell their body what to look like and what to do. A person gets one set of twenty-three chromosomes from their mother and one set from their father. These will determine what color your eyes, skin, and hair will be; whether you'll be a boy or a girl, and so much more." Dr. Apple paused for a moment.

"So when I say Lawana's baby is going to have good genes,

this is what I mean. The chromosomes that this baby has received from Lawana and Mongkut will combine to make a beautiful human being. There is no doubt about it."

Blaine nodded, as Dr. Apple rolled both the map and the blueprint back up and put them back in the bag. She also grabbed a few additional medical supplies from the surrounding shelves and added them to the bag as well. She zipped the bag closed and then looked at the twins, who were still wearing their lab coats and masks.

"Are the two of you ready to do this?"

"Now?" Tracey asked.

"Now," Dr. Apple confirmed.

"But it's next to midnight," Tracey exclaimed, looking at the time on her phone.

"Yes it is," answered Olivia. "What better time than right now to surprise that rotten Nawarak?"

Tracey nodded her head. "Yes, I guess you're right, doc."

"So let's go," the doctor proclaimed. The twins removed their masks and followed the determined doctor out of the supply room.

"This one looks like it'll fit just fine," he thought to himself as he grabbed what he thought was a surgeon's outfit off the poorly lit rack.

"Oh, no, it's pink," he realized. "Well, it'll have to do." He slipped the scrubs on and then added a mask and a pink cap to complete the ensemble. He made sure he secured the cap low and tight on his forehead so his hairless brow was covered. He then slowly opened the door and stepped out into the hallway. He was still fuming that the Expandable Trap Boxes hadn't worked. In function, they'd worked just fine, but those resilient Sassafrases had still managed to escape.

He was still fuming, but he was also encouraged, thinking that this situation here in Bangkok could afford him some good possibilities for stopping the twins. He would look for the perfect opportunity and take it.

Operation Lawana, as Dr. Apple was calling it, was to start by securing an ambulance to drive out to the compound. Olivia Apple was now quickly leading the twins through the hospital's first floor lobby toward the big revolving doors that led out to the drop off/loading zone area at the front of the hospital. She was sure there would be an available ambulance there. The doctor reached the spinning doors first and pushed on them hard as she exited in a rush.

Blaine and Tracey were right on the doctor's heels and tried to be careful not to get caught or smashed by the door's fast, heavy revolutions. They successfully got in, but not out, as the revolving doors slammed to a stop. Both twins smashed face first into the glass doors and then fell down. They looked at each other in a slight daze. Neither one of them had gotten caught in the door. What had happened?

Then, they saw the problem. There, just on the other side of the glass door from where they were, a man who'd been coming into the hospital was pinned in between the previously spinning door and the frame. That man was none other than Nart.

"It's Nart," Tracey cried out. "We've smashed Nart!"

The twins pulled the doors back a little, managing to free

Nart. He dropped to the ground outside. Blaine and Tracey then pushed the revolving doors forward once again and finally made it outside themselves.

"Nart," Blaine proclaimed, leaning down over the fallen man. "We are so sorry we smashed you in the door!"

Nart just looked up and smiled, but it was obvious he had a newly broken arm. Dr. Apple rushed over and saw her hospital friend lying on the ground.

"Oh, Nart," she sighed. "Am I going to have to fix you up every day for the rest of my life?"

"An apple a day," the man responded, laughing gingerly.

Dr. Apple looked at the twins. "Let's just load him up with us. I've already found a willing driver and an ambulance. I can just set and wrap up his arm in the back of the ambulance."

The three helped the accident-prone man back up to his feet and helped him walk over to the ambulance that Olivia had reserved. All four climbed into the back of the vehicle and sat down. Dr. Apple grabbed the back door's handle and started to pull the door shut, but before she could, a hand reached in from outside and held it open.

"Oh, sorry," the doctor said. "I didn't see you there."

A man in a pink nurse's getup hopped into the back of the ambulance, joining them.

"Are you assigned to this ambulance?" Dr. Apple asked the man.

The man in pink simply nodded.

"Well, that's great," Olivia said. "We can use all the help we can get."

She then tapped the window separating them from the driver. The driver quickly put the vehicle into gear, and off they went into the Bangkok night.

Operation Lawana was a go.

CHAPTER 13: OPERATION LAWANA

Lymphatic Rescue

"What did you call him again?" Blaine asked.

"Murse," Dr. Apple replied chuckling. "A male nurse. Combine the words 'male' and 'nurse' and what do you get? Murse."

"Oh now I get it!" Blaine laughed. Tracey and Nart—whose broken arm was now set and bandaged courtesy of Dr. Olivia Apple—were also laughing. However, the man in the pink nurse's outfit was not laughing.

"Oh, get a sense of humor, man," Dr. Apple fussed cheerfully. "We are just joking with you."

She was sitting near to him and she slapped the man on the knee a couple of times to try to get him to lighten up, but it didn't work. The male nurse sat there, just silently stewing. He had his cap low and he was wearing a mask, so that all you could see were his eyes, and those eyes were not happy.

"Okay, Pinky," Dr. Apple gave in, "we will leave you alone. And just let me clarify how I really feel about nurses, male and female; I love them. Nurses do all the hard work but doctors get all the glory. There is no way any doctor could do their jobs effectively without good nurses, and any good hospital around the world would fail if it wasn't for a hard-working nurse staff."

Dr. Apple playfully patted the pink-laden murse one more time on the knee before she switched subjects.

"Okay, people, now for the task at hand: Operation Lawana."

Dr. Apple pulled out the bag that her Thai government friend, Chanarong, had given her, and unzipped it. She pulled out the blueprint and opened it up on the stretcher that was in the back of the ambulance with them.

"The first thing we will do when we reach Kingman's compound is scope it out" the doctor stated. "It looks like there are a few pairs of night vision goggles in the bag here, so we can use them to do that."

Blaine and Tracey looked at each other, excited and a little scared. This was going to be intense.

"After we have a handle on the workings of the compound, we will decide on the best plan of rescue. I don't really think we should go in with guns-a-blazin'; being stealthy is probably our best bet," Dr. Apple looked at her small army that was made up of two kid interns, a gimp, and a murse.

"Does everyone agree?" she asked.

They all nodded.

"If we are successful at finding and rescuing Lawana, we need to be very gentle with her," she added. "Besides being nine months pregnant, I'm also a little worried about her immune system. Interns, you know that the immune system is responsible for defending the body against diseases and that it consists of the immune cells and the organs that produce them. During the last appointment that I had with Lawana, we found out that her white blood cell count was a little low."

"What do white blood cells have to do with the immune system?" asked Blaine.

"They are the key to the immune response" replied Dr. Apple. "White blood cells are produced from the stem cells in the marrow of long bones. They travel in the blood stream and live in the spleen or bone marrow. There are three main types of white blood cells that are frequently used by the immune system. They are neutrophils, macrophages, and lymphocytes."

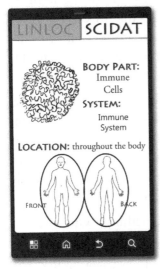

"Neutrophils, macrophages, and lymphocytes?" said Tracey, questioning the funny words.

"Yep, neutrophils, macrophages, and lymphocytes," confirmed Dr. Apple.

"Neutrophils are the germ eaters. These cells travel around the body and look for things that don't belong there. Once they find something, they will eat it up or destroy it. They are the most abundant white blood cells, but they have a very short life. They are the first to respond to an infection in the body, and they are not specific about the bacteria and viruses that they destroy. Macrophages are the cleaners. These are the biggest immune cells in the body. Their job is to clean up damaged cells, bacteria, viruses, and other material that doesn't belong in the body by surrounding and digesting it. They live in the organs, such as in the lung, or they can swim around in the body. Finally, the lymphocytes are the memory specialists of the immune system. There are actually two types of lymphocytes—B-cells and

T-cells—and both of these are formed in the bone marrow. However, the T-cells mature in the thymus gland. These lymphocytes are able to specifically attack certain bacteria and viruses when they enter the body. They can remember previous invaders and destroy them as soon as they enter the body once again." Dr. Apple paused.

"I just don't want to even imagine Lawana having her baby in that maniac's compound!" she shivered. "I want her baby to be born in my clean, pristine women's center. The fewer numbers of germs we have to deal with, the better. Remember that our first line of defense against germs is the skin. Then the organs that can be exposed to outside invaders have a mucus lining to prevent these germs from entering the bloodstream. Saliva and tears also have enzymes that can break down these pathogens. Finally, acid in the stomach kills most of the bacteria and viruses that we eat. So, as you can see, having a low white blood cell count and dealing with lots of germs would not be good, especially for a pregnant woman," Dr. Apple concluded.

The doctor reached down into the gear bag to see what else Chanarong had packed for them. She began placing all the equipment on the stretcher, so that everyone could see what they had. There were the medical supplies that she had added, several rolls of firecrackers, a pair of bolt-cutters, a small plumber's torch, and the aforementioned night vision goggles.

"Well, I'm sure we can think of something to do with this stuff," declared Olivia.

The ambulance continued on through the streets of Bangkok and eventually got to the outskirts of town. Before long, they could see what looked like a small castle glowing in the distance. The driver of the ambulance switched the headlights off and rolled the vehicle to a quiet stop. Dr. Apple looked at her team.

"There are only three pairs of goggles," she announced. "So the interns and I will go scope out the compound. Nart, you and Mr. Murse stay here with the driver and guard the ambulance for now."

Nart nodded but then asked, "Do we have a code word Dr. Apple?"

"A code word? What do you mean?" Olivia responded.

"You know, just in case something goes wrong with our mission, we can shout out this word, to warn each other. Something like Rutherford B. Hayes," the bandaged man clarified.

Dr. Apple looked at Nart with an amused raised eyebrow.

"You want us to shout out 'Rutherford B. Hayes' if something is going wrong?" the doctor asked.

Nart nodded, smiling like a giddy schoolboy.

"Okay, everyone," Dr. Apple informed. "Our code word for Operation Lawana is 'Rutherford B. Hayes,' Please use this word only in emergencies."

She and the twins then put on the high-tech googles and started walking carefully toward Kingman Nawarak's compound.

"Look at this place," Dr. Apple grumbled in disgust. "It really does look like some kind of castle. He really does think that he is a king."

As they got closer, they could see that there was a large wall running around the castle-like building.

"Just like on the blueprint," Dr. Apple confirmed.

The three quietly jogged up and hid themselves in the wall's shadow. They took a second to catch their breath, then they walked the compound wall's entire perimeter, doing reconnaissance. What they found out was that there were quite a few armed guards patrolling around in the courtyard between the wall and the house. Also, by using infrared mode on their goggles, they discovered that there were quite a few people in the house, all of whom seemed to be on the ground level, concentrated in one area. More than likely, these were the poor souls that Kingman had kidnapped. Then, there was one person who was all alone in the very top level, in the biggest room. Presumably, this was Kingman himself. They also found that the wall had two iron gates; a huge front gate complete with a drive, statues, and fountains, and a smaller back gate that was more of just a walkway. Both gates were locked and were being guarded. By the time the three got back to the ambulance, Dr. Apple already had a plan.

"What I had feared is true," she explained. "It looks like Nawarak has kidnapped many more than just Lawana. So, we are going to rescue them all."

This news was a little shocking to the Sassafras twins, only because it seemed impossible, not because they didn't want to attempt it.

"Here is how we're going to do this," the doctor continued. "We will create a diversion at the front gate and then we will sneak all the hostages out of the back gate. Pinky..." Dr. Apple said picking up the rolls of firecrackers and tossing them to the male nurse "...you will hide by the front gate and light these firecrackers at exactly..." she paused and looked at her watch "...12:59 a.m. Nart, you get this," she placed the plumber's torch in Nart's good hand.

"When the guards at the back gate vacate their posts to check on the firecracker sound, use this to cut a hole through the gate's bars."

The good doctor then looked at Blaine and Tracey.

"You will sneak into the compound with me," she directed towards Tracey.

"And you will go back out to the main road with the ambulance driver to secure some additional transportation," she told Blaine.

Dr. Apple took a deep breath, looked at her hodgepodge team, and then looked suddenly worried.

"I am a doctor, not some kind of covert operative. What am I doing? What are we doing? Is this really going to work?"

The male nurse and Nart remained silent, but Blaine and Tracey were both filled with a sudden determination. This was a risk worth taking, an operation worth tackling.

"We can do it!" the twins cried out together. "It is going to work!"

"Okay then!" Dr. Apple vowed, reassurance filling her face. "Let's do this then. Operation Lawana is still a go! Remember, be in your places and ready to go, because snap, crackle, and pop will start us off at precisely 12:59."

Things hadn't played out exactly like he had hoped. Sure, the kids hadn't recognized him yet, but here he was, dressed in pink, being called a murse. On top of that, he was involuntarily teaming up with the two little scoundrels and their friends. He had to figure out a way to capture them or thwart them somehow. He had to stop them from learning science. Maybe he could sabotage this whole Operation Lawana thing. The whole plan did hinge on him, did it not? Maybe he could use the fireworks against the twins and this little team of misfits. Or maybe he should wait. He had seen some syringes full of medicine in the back of the ambulance that he thought would knock the twins out cold if he could just manage to inject them with it. Maybe that was the best plan of action. He looked at his watch. It was 12:57. He needed to make up his mind as to what he was going to do...and quick.

Tracey, Nart, and Dr. Apple were waiting quietly in the shadows by the back gate. Blaine and the ambulance driver were just a little bit off in the distance, waiting with the quietly idling ambulance and a small brigade of motorcycle taxi drivers that Blaine had managed to recruit. It was 12:58, only seconds to go until Boom. Dr. Apple stared at her watch as it clicked to 12:59. All the nervous members of Operation Lawana waited with bated breath for the sound of fireworks. They waited... and waited... all remained silent... nothing sounded.

"What is Pinky doing?" Dr. Apple whispered in frustration. "Is he going to light those firecra —"

The end of the doctor's sentence was cut off by the deafening sound of exploding fireworks. Beautiful colors lit up the night and cracks and pops rang out like gunshots as the rolls of firecrackers, having just been lit somewhere near the front gate, burst on at a frenzied pace.

The three hiding outside the back gate watched through their night vision goggles, as the courtyard guards near the back gate took the bait and left their posts to see what was happening at the front of the compound.

"We're clear!" Dr. Apple declared.

Nart fired up the plumber's torch and prepared to cut through the iron bars of the back gate. However, in doing so, he blinded himself as he looked at the flame in night vision through his goggles. He dropped the torch and it landed on his knee, burning a hole through his pants and scorching his skin.

"Nart!" Dr. Apple whispered frantically. "What are you doing? Take those goggles off! Cut the gate! Hurry, or we'll miss our opportunity!"

After fumbling around for a few more long seconds, Nart managed to get his goggles off, pick the torch back up, and start the cutting process. The little torch worked nicely, and soon a hole big enough to crawl through had been cut out.

Dr. Apple and Tracey bolted into the compound and ran quickly across the courtyard to the house. They flattened themselves up against a wall, took a deep breath, and checked to see if their entrance had gone unnoticed. By the looks of it, it had: There were no guards in sight.

The two females scooted down the wall, to the nearest doorway. They were surprised to find that the door was unlocked and cracked open. They slipped in quickly and quietly. They now found themselves in a short entryway. Dr. Apple led and Tracey carefully followed as the two walked down and out of the entryway, to find a large interior room. Every wall of this rectangular shaped room was lined with tall metal doors, all of which were locked shut with deadbolts.

"This is where Kingman's captives are," Dr. Apple hissed in disgust. "Let's set them free."

Tracey nodded and handed the doctor the bolt-cutters that she had been carrying. The Sassafras girl stood watch as Olivia rushed to the first door and cut the lock off. The doctor quietly opened the door and peeked into the dark room. Sure enough, huddled inside the room in fear, were several women and children. Dr. Apple spoke to them kindly, using some Thai that she knew, telling them that she was there to help and encouraging them to come with her. When they recognized that she was not one of Nawarak's guards, they immediately got up and filed out of the room.

Dr. Apple motioned for them to stay quiet and to go stand next to Tracey. They followed her instructions, as the doctor hustled to the next door with the bolt-cutters ready. These events were repeated over and over again, until Dr. Apple had every door unlocked except for one. There was now a small crowd of freed captives standing next to Tracey, and there was still no sign of guards anywhere.

The fireworks had long since stopped. All the guards must

now be looking for the source and the culprit of the evening's strange events, or had they returned to their posts? Were they even now about to storm into this room and discover Operation Lawana in process? These questions and more ran through Tracey's mind as she nervously waited for Dr. Apple to get the last door open.

The bolt-cutters sliced through the lock, the metal door creaked open, and Dr. Apple let out a squeal.

"Lawana!" Tracey heard the doctor say in way too loud of a whisper.

Tracey watched as a very pregnant Thai lady hobbled out of the dark room and met Dr. Apple in a tearful embrace.

"Olivia!" the woman cried. "How did you find me here?"

"That doesn't matter right now," the doctor responded, smiling and crying. "Let's just get you and everyone else out of here safely."

Lawana nodded as she and Olivia joined Tracey and the group. Dr. Apple gave the signal to the Sassafras girl, and Tracey tiptoed out of the room into the short entryway, leading all the scared women and children, she hoped, to what would soon be their freedom. Tracey reached the door that they would exit from and cracked it open to peek out into the courtyard. Oh no! One of the guards had returned, and he was walking directly towards the door. Tracey yelped in fear. The guard heard her and then spotted her. His face showed surprise that quickly turned to anger, as he burst into a run to come and get her. Tracey was frozen. She couldn't move, and no shouts of warning were coming to her mouth. The guard's sprint got him to the door in seconds. He reached out a hand to grab the door and pull it wide open, but just as he did, a crutch came swinging out of nowhere and walloped the guard right in the face.

The guard crumpled to his knees and slowly fell to his side, completely unconscious, as Nart stepped around the door frame, holding a crutch and sporting a goofy grin. The casted and bandaged man opened the door up wide for the girl and showed her that the coast was clear.

At last, Tracey stepped outside and motioned for the train of captives to quietly follow her. The twelve-year-old led the fast moving single-file line as they made a beeline across the courtyard to the back gate. Tracey dove through the hole in the gate and then kneeled in a position enabling her to help the others through. The women and children ducked through the cut gate, one by one, leaving Kingman Nawarak's compound and entering the shadows of freedom.

Tracey helped the pregnant Lawana through, next came Dr. Apple, and then last but not least, Nart came tumbling through the make-shift exit. Blaine was right there to greet the freed captives and lead them carefully to the waiting motorcycle taxis.

Speedy Germs

Operation Lawana miraculously continued to proceed according to plan, unnoticed by any of Kingman Nawarak's guards. The fireworks diversion had really thrown them for a loop. There were thirty-six freed captives in all, which Blaine managed to get successfully saddled up on the motorbikes in record time.

There weren't quite as many motorcycles as people, so some of the bikes were carrying more than one extra passenger, but the confident Thai motorcycle taxi drivers didn't care. They were just glad that they could help. They all despised Kingman Nawarak and thought he was an embarrassment to the great country of Thailand itself.

Once the captives were all loaded up on their bikes, the motorcycle taxi drivers put on their helmets, tightened their reflective orange vests, made sure their flip-flops were secured between their first and second toes, and then waited for the signal to go.

Blaine had saved the spot in the ambulance for Lawana. The Sassafras boy now opened up the rescue vehicle's back doors and helped the pregnant woman up into a comfortable seat. The male nurse had managed to make it back to the ambulance without getting caught, and he was already seated in the back of the vehicle. Dr. Apple and Tracey jumped in and joined Lawana and the nurse.

"Wait a second!" exclaimed Dr. Apple. "Where's Nart?"

Suddenly, they heard something they hadn't wanted to hear.

"Rutherford B. Hayes! Rutherford B. Hayes!" The code word rang out in the night, being shouted by an alarmed Nart as he lumbered quickly toward the ambulance on his crutches.

"Rutherford B. Hayes!" he shouted again.

"OK, OK!" Dr. Apple protested, as she pulled the man up into the back of the ambulance. "We know something is going wrong with the mission, but what is going wrong?"

Nart pointed out the open back doors of the ambulance back toward Kingman's compound.

"They're onto us!" the man pointed, with a horrified look in his eyes.

Everyone looked and, to their terror, saw angry guards spilling out of the back gate and running straight toward them.

"Go! Go!" Dr. Apple shouted in urgency, to no one in particular and everyone all at the same time.

Blaine, who was still standing just outside the ambulance, slammed the back doors shut, securing his friends inside. Then he jumped onto the back of a motorcycle taxi he had saved for himself. All the motorcycle taxi drivers and the ambulance driver took that as their signal, and they simultaneously fired up their engines and took off.

Nawarak's guards were caught in a cloud of dust as all the escaping vehicles accelerated away. The escapees bolted toward the road that led back to the city, with the ambulance leading. Once on the road, Kingman Nawarak's compound became smaller and smaller behind them as they sped toward Bangkok's interior.

Blaine used the motorcycle's mirrors to look back at the shrinking image of the castle-like building. He was more than happy to be leaving it behind, but wait! What was that he saw in the reflection? A whole drove of something loud, fast, and black was streaming out of the front gate. The guards were giving chase on motorcycles of their own. Blaine felt a collective surge in velocity as the ambulance and all the motorcycle taxis notched up their speed to an even higher gear, aware that they were being chased.

Nawarak's army of flashy black motorcycles was also joined by a totally black truck with dark tinted windows. The truck brought up the rear of the pack. Blaine didn't remember how long it had taken them to get out here from the hospital, but he was sure even with their speed and head-start, Kingman's men were going to catch them.

He had beaten everyone back to the ambulance after he had started the fireworks. He had managed to grab a couple of the loaded syringes without it being detected. He held the syringes, even now, in his hand, which was tucked into the front pocket of his pink scrubs. The bad thing was that only one of the twins had gotten into the ambulance. The other one was on a motorcycle outside.

"Oh, well," he thought. "One twin is better than none. I'll just have to think of something else for the boy."

He would try to stick the girl with the needle quickly when no one was looking. This was going to be easier said than done, considering how packed the back of this ambulance was right now, but he had already decided that this was the best plan. The doctor was looking out the ambulance's back windows, saying something about Kingman Nawarak's men chasing them with a truck and motorcycles. He didn't much care about any of that. His sole focus was taking down the Sassafrases.

Blaine was surprised when they managed to make it back to the city without being caught by the black fleet behind them. He was also surprised how jam-packed the streets of Bangkok still were this late at night. Didn't anyone sleep in this city? This wasn't good. All this traffic would surely aid Nawarak's pursuing motorcycle henchmen in catching them. As they hit the gridlock of cars, buses, and brightly colored taxis, the ambulance driver clicked on his lights and sirens trying to clear a path.

Blaine looked back to see that none of the motorcycle drivers were slowing down at all as they hit the traffic. As a matter of fact, they seemed to be speeding up. Blaine's heart jumped in his throat as his motorcycle driver zipped into a narrow gap between two cars. He was sure they were going to crash or at least get scraped up, but no contact was made. The driver swiftly and skillfully guided the bike through the two cars and continued onward up and around the slow-moving traffic. Blaine looked to his left and right to see the other motorcycle taxi drivers were doing the same. They were all treating this gridlock like it was a wide-open multi-lane highway.

Blaine looked behind them again, thinking that surely they could escape easily now. However, Kingman's motorcycle men weren't slowing down either. It was a full speed chase through the crowded neon nighttime streets of Bangkok. The growling roar of the motorcycles' engines echoed loudly off the buildings as the dozens of bikes ripped by.

Blaine held on for dear life as the motorcycle he was on dodged and weaved, churned and turned, sometimes with only centimeters of clearance on either side. He made sure his legs were pulled in close to the motorbike so they wouldn't hit anything.

Blaine looked ahead and saw a row of buses blocking their path. There were gaps in between the buses, but those gaps were way too small for their motorcycle to fit through. The driver wouldn't dare try it, would he? Blaine didn't think a paper airplane could fit through the gaps between those buses, much less a motorbike.

The driver accelerated toward the buses, aiming for one of the tiny gaps. At the last second, the driver veered the bike and took the only available pathway, which just so happened to be the sidewalk. Blaine now found himself on the back of a motorcycle that was speeding down a walkway crowded with people and vendors selling merchandise from street stalls. The vendors' eyes got big and the pedestrians dove out of the way as the motorbike zipped by.

The black motorcycles continued to follow. They weren't seeing the gridlocked streets or busy sidewalks as obstacles either. One of the black motorcycles had singled out the bike Blaine was on and was now trailing him and his driver on the sidewalk. He was gaining on them with every rotation of the tire. Blaine peeked around his driver to look ahead and saw that the sidewalk soon dead-ended into a small canal. The sidewalk they were on had no bridge—it simply dropped off into the water. Instead of slowing down, though, the motorcycle driver pulled back on the throttle, bringing his bike to full speed.

What he knew—that Blaine didn't—was that there was a sudden incline at the end of the sidewalk that could send them airborne. The motorcycle hit that incline at full speed, shot up into the air, and easily sailed safely over the canal.

The black-clad motorcycle driver behind them saw the canal but not the incline. He slammed on his bike's brakes but not in enough time to come to a stop. Blaine watched as the black motorcycle skidded over the edge and down into the waterway. When he looked back in front of them, he saw that they were now right behind Nawarak's black truck, which was chasing the ambulance his sister and friends were inside.

This was his opportunity. The doctor was standing up and pointing out the window, showing everyone how close the black truck that was chasing them had gotten to their back bumper. Everyone was staring out the back windows and nobody was paying any attention to him, and the girl twin just happened to be standing right in front of him.

He gripped the syringe tightly and pulled it out of his pocket. He lifted his hand up and then brought it down quickly. But just as he did, the bandaged one called Nart jumped up right in front of him for a better view of the back windows. The needle sunk squarely into the back left shoulder of the misfortunate man.

The man looked back at him with a look on his face that was asking, "What just happened to me?" But almost immediately, the man's eyes rolled back and he fell unconscious, crashing down onto the stretcher that was the centerpiece of the back of this ambulance. At the precise moment this happened, the black truck rammed into the back of their vehicle. He felt himself pitching forward. He fell and landed right on top of the unconscious Nart. Then, to make matters worse, the back doors of the ambulance flew open and the stretcher rolled right out.

Blaine watched as the black truck rammed into the back of the ambulance, causing the vehicle's back doors to fly open. The black truck then swerved as if avoiding something. Blaine didn't know why until he saw that the stretcher had rolled out of the back of the ambulance. Since the truck had swerved, Blaine and his driver were now the ones who were on a crash course with the runaway stretcher.

The driver leaned just in time, causing their motorcycle to miss the stretcher by mere inches. But then Blaine saw the escaped stretcher had two people on it: the murse and Nart. If he let the stretcher just roll on, it would surely get crushed by oncoming traffic.

What Blaine did next seemed to happen in slow motion. As the stretcher rolled past them, the Sassafras boy reached out and grabbed it just before it got away. This jerked Blaine's body sideways and backwards, threatening to pull him off the motorcycle. However, the quick-thinking motorcycle driver reached back and caught Blaine before he fell.

Now, the twelve-year-old found himself lying stomach-down on the back of the bike, with his legs wrapped around the driver's waist and his hands were holding onto the rolling stretcher. The murse hadn't been as lucky as Blaine. All the changes in movement and inertia had thrown him off the stretcher and onto the hood of a taxi behind them.

The taxi driver, stunned by the sudden appearance of a person on his hood, jerked his car off the main road and disappeared onto a small side street. Blaine had lost Pinky, but he wasn't going to lose Nart. The accident-prone man was still on the stretcher and, apparently, was unconscious. Blaine wrapped his hands around the bars of the stretcher as tightly as he could.

The black truck had disappeared but the band of black motorcycles pursued on. They were determined not to let their ex-captives get away. Blaine heard a strange noise. He looked down to see that the quickly circulating steel wheels of the stretcher were causing sparks to fly up from the road as they raced on. The sparking, speeding stretcher was making it harder to weave in and out of traffic, but still the driver was managing to do just that. Blaine hoped the driver wouldn't have to do this much longer. They had to figure out a way to lose these black motorcycles!

Back in the ambulance, Dr. Apple managed to pull the back doors closed again. She plopped down to a seat, trying to gather her thoughts. How could they get away?

"We've got three dozen women and children out there zipping through Bangkok traffic on the backs of motorcycle taxis," she shared with Tracey and Lawana. "We have an intern desperately clinging on to a rogue stretcher. We ourselves are riding in an ambulance with a smashed bumper, and those black motorcycles are still chasing us. Can things get any crazier?"

Just then, Lawana waved her hand to get Olivia's attention. The Thai woman's face was wincing in pain.

"The baby!" she grunted. "I think it is coming!"

"I shouldn't have asked that last question!" Dr. Apple declared, seemingly laughing and crying at the same time.

Tracey suddenly had a great idea. She slid open the glass window separating them from the driver and asked him if he would pass her the vehicle's megaphone. The driver obliged and handed the girl the device that would enable everyone outside to hear her voice. Without even asking Dr. Apple's permission, Tracey put her plan into action. She pushed the appropriate button and began speaking into the megaphone.

"People of Bangkok," she shouted out. "My name is Tracey Sassafras. I'm not Thai, but I love Thai people, and that's why I'm in the back of this ambulance. My friends and I are attempting to rescue Thai women and children that have been kidnapped by Kingman Nawarak. Please help us! The riders on the black motorcycles are Kingman's men. They're chasing us so they can catch us and take these poor people into captivity! People of Bangkok, please help us!"

Vendors, pedestrians, and other drivers heard her message

loud and clear. Everyone in Bangkok knew who Kingman Nawarak was, and nobody liked him. People began looking around angrily, trying to see the black motorcycles that they'd heard the girl's voice speak of. The people of Bangkok were definitely ready to help.

Tracey repeated her message over and over again in the megaphone, making more and more people aware of the menacing black motorcycles. Her idea turned the tide, as vendors began smacking the black motorcycle drivers with different items as they raced by. Pedestrians tackled them off their motorbikes and drivers of different vehicles would open their doors just as the menacing motorcycles were coming by, causing their drivers to be flipped off. One by one, Kingman Nawarak's motorcycle driving henchmen were put out of commission.

When it was clear they were no longer being pursued, Tracey handed the megaphone back through the sliding window to the driver, and then turned back to Dr. Apple and Lawana.

"We did it!" Tracey rejoiced. "Operation Lawana was a success!"

Dr. Apple looked relieved but not convinced. "I'll call the operation a success when this baby is successfully delivered." Lawana groaned in pain as she continued to experience the contractions of labor.

"Lawana, I do believe you're right," Dr. Apple soothed as she washed her hands with some disinfecting solution. "I think this baby is coming right now. I'm a little worried about your white blood cell count and any germs we might encounter during this delivery, but we can prevent a lot of germs simply by washing our hands."

Lawana's face showed she didn't that care about that right now. She just wanted to have the baby and be out of pain.

"Just breathe, friend," the doctor demonstrated some calm breaths.

The traffic light that they had momentarily stopped at turned green and the driver moved forward. But just as he did, the black truck came racing into the intersection headed straight for them. The ambulance driver swerved, preventing a full-on collision, but the truck still hit and ricocheted off of them with great force. The smoking and battered ambulance spun off into a side street as the smashed-up black truck disappeared in the opposite direction.

The ambulance driver guided the clanking vehicle into an alley where the poor ambulance sputtered and died. Dr. Apple was opening up the ambulance's back doors to see exactly where they were just as Blaine's motorcycle pulled up.

"Oh, great," the doctor exclaimed when she saw where they'd come to a stop.

"A dead end alley with nothing but rats and trash. What an ideal place to deliver a baby," she raved sarcastically. "The germs! Oh, the germs that are in this place!"

Tracey was pretty sure Dr. Apple was on the edge of losing it.

"Tracey, you know that germs are bacteria, viruses, fungi, or protozoa that enter the body and cause it to become sick. They're normally microscopic so that we can't see them as they sneak into our bodies. We don't know we've been attacked until we begin having the symptoms of the sickness they cause. Remember that bacteria are single cell organisms that get all their nutrients from the environment. They can reproduce within the body and cause infection, but not all bacteria are bad. Some bacteria are beneficial to us; for example, the ones that help us to digest

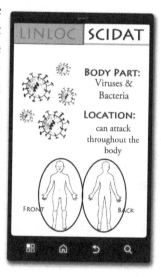

our food." Dr. Apple paused for a breath as Blaine approached the back of the ambulance.

"Viruses, on the other hand are single-celled organisms that need to live in a host cell to reproduce. They first attack a cell from our body. Then, they break through the cell wall barrier and invade the cell. The virus then takes charge of the cell and multiplies by using the cell to produce copies of the virus. Finally, the cell is destroyed and viruses disperse into the body. We need to keep in mind during this delivery that viruses are easily spread through contact and through the air. The last things we need to worry about are protozoa, which are single-cell organisms that love moisture. They can cause diseases that are typically spread through water." As Dr. Apple finished her fast-paced introduction of germs, she felt a gentle touch to her arm. She looked and saw that it was Tracey.

"Dr. Apple, you are an amazing doctor," the girl reassured. "Germs or no germs, your friend needs you now. She needs an Apple today.

Dr. Apple looked toward her friend, who was lying in the back of the ambulance on the floor. Lawana was smiling and nodding in agreement. The anxiety on Olivia's face left, and her normal confidence returned.

"An apple a day keeps the doctor away. That's what they say, anyway, but hey, an Apple is your doctor today!" Dr. Apple smiled. "Let's get you on the stretcher and deliver this baby."

Olivia climbed back into the ambulance to assist her friend as she moved out of the way of the stretcher. The twins moved the rolling bed back into the ambulance and rolled the unconscious Nart onto the bench. Then, they helped Lawana as she laid down on the stretcher. Just as she did, a beaten-up, black truck slowly pulled into the alley.

CHAPTER 14: BACK TO SUMMER'S LAB

Cells

The door opened up and a jeweled boot stepped out and hit the pavement. A man wearing a designer suit appeared, stepping out from behind the truck's black door. The big sneering male was draped in jewelry from head to toe. Tracey saw he even had some gold teeth, when he began arrogantly laughing.

"Looks like I've got you now, doesn't it?" He slowly walked toward the open back doors of the ambulance, his expensive shoes clapping on the alley's dark surface.

"Kingman Nawarak?" Dr. Apple asked.

The man simply nodded. He stopped a few feet from the ambulance and peered past Dr. Apple and the twins, to see the laboring Lawana.

"It looks like you've got something that belongs to me," he sneered.

A look of resolve once again filled Olivia Apple's face. Though the seemingly untouchable man in front of her was violent, lawless, and powerful, his statement about Lawana had the good doctor fuming.

"She does NOT belong to you!" Dr Apple stated boldly, with no trace of intimidation in her voice.

"Oh, she doesn't, does she? I beg to differ with you on that, doctor," Nawarak responded, with an arrogant chuckle.

He took a step closer, but Dr. Apple stood her ground.

"Everything in this city belongs to me," the richly dressed man emphasized. "If I want it, I take it. Bangkok is the city of Kingman Nawarak."

"Bangkok is not your city!" Dr. Apple retorted. "Bangkok belongs to the real king and to the people of Thailand. You are nothing but a disgusting, self-absorbed maniac that wears too much jewelry."

At that, Nawarak ruffled the pile of gold chains around his neck and raised his hands into the air, proclaiming to the whole city. "I am Kingman Nawarak! And this IS my city! Bangkok, I will do with you as I wish. I will continue to take all your money. I will continue to build up my luxurious castle on the edge of the city. I will continue to kidnap your women and children! That is who I am, that is what I do, and nobody can stop me!"

Just as Kingman finished his exclamation, half a dozen men dressed in all black with helmets, came reppelling down into the alley on ropes strung from the tops of the buildings on either side. Before they even hit the ground, it was apparent that these men were uniformed Thai military and that they were there to capture Nawarak. All six seemed to be running towards Kingman before their boots even hit the ground.

THE SASSAFRAS SCIENCE ADVENTURES

The jewel-laden man froze in fear as he realized what was happening. The six Thai men tackled Kingman Nawarak to the ground and tied him up and handcuffed him so quickly that he had no chance to run. One of the military men then stood up and took his head gear off; it was Chanarong.

"Chanarong!" Dr. Apple exclaimed. "Where did you…? How did you…? I am so glad you are here!"

She ran up and gave her friend a grateful hug. "Thank you! You rescued us."

"No, Olivia," Chanarong said. "You rescued us from Kingman Nawarak. It is we, the people of Bangkok, that need to thank you. Fearlessly, you went out to his compound and you saved all the women and children he'd kidnapped. And after a harrowing chase back to the city, you lured the man to this alley, where he just made a clear confession of guilt. We recorded the confession, we captured the man, and now, thanks to you, the terrible reign of Kingman Nawarak is over!"

Dr. Apple, Blaine, and Tracey looked over to where Kingman was lying on the ground, tied up and defeated.

"Operation Lawana was a success!" Tracey exclaimed.

Dr. Apple nodded her head and smiled.

"Speaking of Lawana," the pregnant woman moaned from the back of the ambulance, "I could use some help right now."

"Oh, Lawana!" Dr. Apple apologized, turning and rushing back toward the ambulance.

"C'mon, interns," she added to the twins whom she still thought were soon-to-be doctors. "It's time to bring this baby into the world!"

THE SASSAFRAS SCIENCE ADVENTURES

"No way!" Blaine said, excited. "We get to go back and see Summer Beach again!"

"That's awesome!" Tracey exclaimed. "I really liked her. I just hope we won't have to live through another heliquickter crash this time."

"Yeah, that's for sure," Blaine agreed.

The Sassafras twins sat in a quiet corner of the hospital looking at the LINLOC apps on their smart phones. They had just successfully entered SCIDAT data about reproduction, DNA, white blood cells, and germs. It was quite a chore to remember all they had learned from Dr. Apple on these subjects, and then get it texted into their phones. But, they continued to prove their ability to do this, location after location, and subject after subject. They had also flipped through the archive app and had chosen the appropriate pictures to send in with their SCIDAT information. It took a little time to do this, but it did take away the pressure of having to take pictures as they went.

Bangkok had been good to the twins. They had learned more about the human body, and had also been a part of helping to rescue some Thai women and children. Kingman Nawarak had been taken into custody by Chanarong and his task force. All the motorcycle taxis had made it back safely to the hospital with their rescued passengers.

Thankfully, they hadn't seen the Man with No Eyebrows at this location, and to the joy of everyone, Lawana had given birth to a baby boy. Despite all of Dr. Apple's worries about Lawana's low white blood cell count and all the possible pathogens floating around in the alley, the delivery had gone fine. Lawana and the baby were completely healthy in every way. She named him Mongkut after his father.

Really, the only thing that had gone wrong the whole night, other than banging up Nart a bit, was that they had lost the male nurse. After he'd flipped off the runaway stretcher and landed on the hood of the taxi, no one had heard from him or seen him again. Dr. Apple told the twins not to worry about that. She said that she would work on finding him.

So, here they were now, in the wee hours of the morning, preparing to zip off to the arctic tundra in Alaska: Longitude 163° 4' W, Latitude 63° 3' N.

"The subjects we'll be studying this time with Summer are cells, hormones, and glands," Tracey shared.

"Cells?" Blaine asked. "I thought we just learned all there was to learn about cells when Dr. Apple taught us about the white blood cells."

"I guess not," Tracey responded. "I guess there's more to cells than we know about yet."

"Well, let's go find out!" Blaine exclaimed as he finished cinching up his harness and calibrated his carabiner.

Tracey's harness was already on and fastened, and she was also setting her carabiner to the correct coordinates. A few moments later, the Sassafras twins were zipping from the Pacific Rim to the Alaskan tundra at the speed of light. They landed with a jerk, and when their senses normalized, they saw they'd landed in a familiar frozen field—the same one they had landed in separately the first time they'd come to Alaska.

There was a little less snow this time, but the large, flat, tree-and-mountain bordered field was still mostly white. As they were both expecting, the ground opened up and down they went, on the spiral slide, down into Summer Beach's high-tech underground lab. As soon as the twins hit the neatly padded landing area, they heard a high-pitched scream of delight.

"Aahhh! Blaine and Tracey Sassafras!" Summer's familiar voice squealed. "You're back! Oh, I'm so glad you're back!!"

"We're glad to be back," the twins said in unison.

The blond, frizzy-haired, high-energy scientist ran over to greet the twins with outstretched arms, her lab coat flapping behind her. She grabbed them both at once, with hugs and giggles, and began dancing around with them in an exuberant circle. Blaine and Tracey had known what to expect with Summer Beach this time around, but they both still felt a little awkward doing the happy dance with the enthusiastic scientist.

"Sweet sinapis and porcina on rye, you two Sassafrases are rocking my face off with scientific awesomeness!" Summer shouted in joy. "You successfully completed your zoology studies, and now you're flying through anatomy like old pros!"

"What are sinapis and porcina on rye?" Blaine asked.

"Oh, that's Latin for mustard and ham. I really like sandwiches," Summer giggled.

The female scientist let go of the twins and hopped over to one of the floor-to-ceiling translucent data screens. As the twins followed her, they looked around the lab. Though it had only been a week or so since they'd been here, they'd already forgotten how nice and neat Summer's workshop was. Everything was sleek and clean, and the big circular shaped room had all kinds of cool scientific specimens displayed in colorful, liquid-filled tubes or beautifully lit boxes encased in the walls. Last time they were here, Summer had only covered musk oxen, snow geese, polar bears, and mountain goats, but the twins could tell just by looking around that if she or Uncle Cecil planned to, Summer could teach them about almost anything relating to science.

Summer was moving her fingers around on the see-through screen, flying through different pages, but she stopped when she got to the tracking screen, "I've been watching your progress, you cute little Sassafras twinkies. And I continue to be impressed. Since you left me, you've not only gone to South Georgia Island and back to Dreamy McDreamerson's, or rather Cecil Sassafras's house, you've also been to five different continents, getting the scoop on anatomy. Now, I, Summer T. Beach, get to be one of your local experts— AGAIN! I'm so excited I could do a back handspring! As a matter of fact, I will!"

The hyper scientist took a couple of running steps and completed a perfect handspring. Upon landing the move successfully, she clasped her hands in delight and looked up.

"Oh, I'm so glad that Cecil asked me to be a part of this section of your studies, too! I relish the thought of continuing to be a part of your super summer scientific zip lining adventures!"

Summer flipped through some pages on the translucent data screen until she got to a page about cells. "Just like before," she continued. "Here in my lab, you can give your texting fingers a rest because I can upload all the SCIDAT data straight to your phones. Of course, we will still talk about the information and you'll need to get pictures of all the topics."

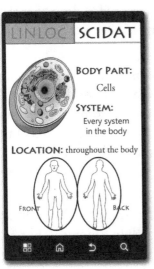

Without hesitating, Summer began reading what was displayed on the screen about cells: "Cells are tiny living units. They are typically microscopic, but some can be seen with a magnifying glass. There are two main types of cells: prokaryotic and eukaryotic. Prokaryotic cells include bacteria, which you Sassafrases have already learned a bit about. Eukaryotic cells include plant and animal cells. Both plant and animal cells contain a nucleus and several organelles. They both also have membranes, but the inside of a plant cell is under pressure, so their cells have cell walls. Plant cells also contain chloroplasts, while animal cells don't."

Summer paused and looked at the twins, smiling. "This is all so very exciting, isn't it?"

The Sassafrases nodded.

Summer went on, "Now, back to those two kinds of cells. Prokaryotic cells have no nucleus, but they do have a membrane. They are very simple cells and are usually found by themselves. Eukaryotic cells, however, are usually found as an organized part of a bigger system and have a nucleus. They are quite a bit more complex, which means that they can carry out specific functions and have multiple structures inside the cell.

"The human body is made up of millions and millions of eukaryotic cells, each with its own shape, size, and function. For example, muscle cells, which are also known as muscle fibers, are long and striped cells that can contract. Liver cells are cube-shaped which allows blood to flow between them. Red blood cells are shaped a little like doughnuts without a hole, so they can easily move through the blood vessels and carry oxygen to other cells. While, nerve cells have long branches off their main cell body so that they can send messages from one cell to another." Summer paused for a breath.

"Cells make up tissues; tissues make up organs; organs work together to form a system; and the systems work together to form the body." Summer finished reading the information that was written on the screen and then clicked on a tab at the bottom of the page that said, "Upload." The twins immediately felt their phones vibrate. The SCIDAT data that they needed on cells was now ready to go in their phones.

The scientist turned to the twins as she said, "We will talk about the specific parts of a human body cell in a moment, but first I have a surprise for you."

The scientist clapped her hands and smiled her biggest smile yet. She flipped through a couple of pages on the translucent data screen until she got to a page that read, "Microscope application for zip line smartphones."

"Ulysses S. Grant and I, in collaboration with your handsome uncle and his fine prairie dog, have invented another app for you guys to use on the adventure trail."

"A microscope app?" Tracey asked.

"Yes indeedy!" Summer responded exuberantly. "A microscope app! Let me upload it to your phones real quick and then I'll show you how it works!"

Summer pushed the upload tab on the screen, and after a few seconds, a new icon appeared on the screens of the Sassafrases' phones, showing they now had the new microscope apps. Summer took hold of the twins' hands and led them over to one of the labs nearby work stations.

"I've set out some slides on this table," she informed. "Normally, what any scientist would do is grab these slides, stick them under a microscope, and see what they hold. But now, with your new microscope applications, you two can do this work with your phones!"

Summer grabbed a couple of the slides and placed one in front of each twin.

"Go ahead! Try it out!" she encouraged excitedly, watching to see if the twins would dig the new app.

The Sassafrases both turned on their new microscope applications, then placed their phones' camera lenses over the slides. Immediately, they saw clearly magnified images of the very cells that Summer had just been talking about.

"Now, just snap a picture and you can send it in later for SCIDAT!" Summer declared.

"This is so awesome!" the twins exclaimed, enthusiastically, at the exact same time.

Not able to contain herself over the twins' excitement, Summer grabbed Blaine and Tracey and did one of her patented happy jumping hug dances.

Inside the Cell

He had scraped his right side up pretty well when he'd jumped off the hood of the taxi, but it was nothing that wouldn't heal. What had hurt him more was the fact that he had, yet again, failed to stop those Sassafras twins. However, each time he failed, it didn't get him more discouraged. It only fueled his determination to stop them all the more. He WOULD stop those twins and, in doing so, he would crush the dreams of his archenemy, Cecil Sassafras.

With his well-placed hidden cameras in Cecil's house, he had the ongoing ability to track the twins' whereabouts. He knew they were now in Alaska on their second trip to see Summer Beach. He hadn't bothered them there the first time, but that wouldn't be so this time.

He had spied out the longitude and latitude coordinates that had landed the twins on the field directly over Summer Beach's underground science lab. He didn't know how he could get from the field down into the lab. So he tweaked the LINLOC coordinates just a little bit, enabling himself to land inside the underground lab.

He wasn't too worried about changing the coordinates. He did this at virtually every location, but he was a little worried about the availability of good places to hide once he did land in the lab. He knew Summer Beach ran a much cleaner lab than Cecil Sassafras did, so there probably wouldn't be any big piles of junk to hide behind.

"Oh, well," he thought to himself. "It is worth taking the risk. I'm sure I can figure out a way to stay hidden. Then, I can definitely come up with another brilliant sabotage plan."

He cinched up his harness, turned the three rings on the carabiner to the correct coordinates and let it snap shut. In a matter of seconds, off he went, at the speed of light.

"Is he hiding?" Tracey asked.

"No, he's not hiding," Summer Beach answered.

"Then why haven't we seen him yet?" Blaine asked.

"Because, he's putting the finishing touches on a surprise he has for you!" Summer shared, jumping and clapping.

"Another surprise?" the twins asked. "Ulysses S. Grant, the arctic ground squirrel, has a surprise for us?"

"Yes-o-yes!" Summer squealed and smiled like she already knew what the surprise was but would never tell. "He definitely has a surprise for you. And it's a good one! You know, besides being my lab assistant, he is also an inventor extraordinaire. So you never know what he's going to come up with. I have to say that the surprise he has waiting for you today may just be his coolest idea yet."

The Sassafras twins' minds began racing with all the possible things that Ulysses' surprise could be.

"But before we get to the squirrel's surprise, let's talk about organelles." Summer's voice brought them back to reality.

"Organelles?" Blaine asked.

"Yes, organelles," Summer confirmed, as she began skipping back over to the floor-to-ceiling data screens. "They are the components of the human body cell."

The twins followed the scientist over to the screens and listened as she found the correct page and began reading.

"The cells of the human body can look externally different, but internally, there are a lot of similarities. The structures found inside of a cell are called organelles, which means 'tiny organ.' Each of these tiny organs helps to keep the cell working properly. They can only be seen with a microscope and some of them can only be seen with a very powerful microscope called an electron microscope, which can magnify an object up to millions of times."

Summer looked away from the screen for a second toward the twins with her usual smile. "I'm proud to say that the microscope application on your phones is 'electron' caliber, so right after I finish reading this data and we get it uploaded, we will dance back over to that table where the slides are and check out every single organelle that I'm about to tell you about."

The twins yawned. Their new app was cool; this science was cool; Summer Beach and her arctic ground squirrel were cool. Everything was cool, but they were tired. Summer didn't see them yawn because she'd already started reading the data again.

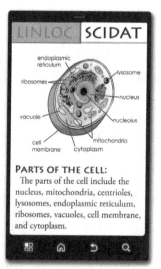

"The following organelles are common to almost all the cells in the human body: the nucleus, the mitochondria, the centrioles, the lysosomes, the endoplasmic reticulum, the ribosomes, the vacuoles, the cell membrane, and the cytoplasm.

"The nucleus of the cell contains

the genes and controls cell function. The mitochondria make the energy for the cell from food that we eat. The centrioles help cells to divide and the lysosomes contain helpful enzymes that the cells need. The endoplasmic reticulum helps make and store the chemicals that are necessary for life. While ribosomes act as the cells' messengers by carrying those chemicals to different parts of the cell. The vacuoles store liquid or gas in the cell. The cell membrane acts as a barrier for the cell, and finally, cytoplasm is a jelly-like substance in the cell that holds all the organelles." Summer let out a sigh of satisfaction as she finished the information.

She tapped the upload button at the bottom of the screen. The twins nodded and yawned as their phones vibrated, confirming they had just successfully uploaded the data. Then, just as she said she would, Summer literally danced her way back over to the work station, where more slides holding cells awaited. The twins slowly followed the scientist to the table.

Summer set out a row of slides along the table. "OK, little Sassy Frassies," Summer said joyfully. "Just go down the line and check out all these slides using your microscope apps."

The twins happily but groggily followed their scientist friend's instructions. Tracey was ahead of Blaine, looking at each slide first. Blaine noticed that at each slide, Tracey's face was getting closer and closer to her phone. When she reached the last slide, her face actually rested on the screen of her phone. Blaine looked at his sister curiously.

"Tracey?" he asked. "What are you doing?"

Tracey just remained frozen like a statue in her comical position, leaning over the table with her face on her phone and her hands by her side.

"Tracey?" Blaine tried again. "Are you...asleep?"

"Indeed, I think she is," Summer laughed. "It looks like an extreme case of sonic lag."

"Sonic lag?" Blaine asked, scrunching up his face trying to remember the term. "Oh, yeah, sonic lag. It's the zip lines messing with our inner clocks or something, right?"

"Right," Summer confirmed.

The scientist stepped over and gently woke up the Sassafras girl from her standing slumber.

"Where did that... the what... the why?" Tracey blurted as her eyes slowly opened back up.

"Sonic lag," Blaine told his sister. "You got a bad case, Trace."

"Well, it makes perfect sense that the two of you would be incredibly tired at this point in your journey," Summer stated. "This is basically the fourth time you've lived this day."

"What?" the twins asked, totally clueless.

"You haven't slept since you were in Beijing at the Bird's Nest," the scientist explained.

"We haven't?" the twins asked.

"Nope, you haven't," Summer responded. "You two got a good night's sleep at the Bird's Nest, and then you spent the next day at the Great Wall. That evening, you zipped back to that morning, landing in Lubbock, Texas, where you only spent two or three hours. Then, you zipped from that morning back to that night in Bangkok, Thailand. You made it to the early hours of tomorrow in Bangkok, but then, when you zipped here to Alaska, you zipped back to today, the fourth today that you've lived."

The twins still didn't understand.

"So we haven't slept in one day? Or we haven't slept in four days?" Tracey asked.

"Yes," Summer answered.

The scientist walked over to the wall and pushed a couple of buttons, opening up two egg-shaped doors, revealing something the twins immediately recognized and appreciated.

"The sleeping pods," both the Sassafrases said, in a sort of half statement, half sigh.

"Yes, the sleeping pods," Summer smiled. "No better way to fight off sonic lag than to get a good long rest in a comfy, cozy, Ulysses S. Grant designed sleeping pod."

Miss Beach then pointed at the smartphones the twins had left lying on the workstation table. "Don't worry about your SCIDAT data. We'll finish that whenever you guys wake up."

The twins hadn't realized they had been awake for such a long time. The Bird's Nest seemed like forever ago and now this phenomenon that Summer was calling sonic lag was hitting them like a ton of bricks. It was all they could do just to keep their eyes open long enough to walk across the room and climb into the sleeping pods.

As they both climbed into their bed-sized pods, the soft cushions and pillows immediately embraced their weariness.

"Nighty night, you two," Summer whispered.

She reached up to push the buttons to close the pods, but as she did, a bigger egg-shaped door began slowly opening up. Ulysses S. Grant, the arctic ground squirrel, scurried in excitedly. The twins looked at the friendly lab assistant through slouching eyelids and saw that he wasn't alone. He was joined by a whole crowd of...

"Are those...?" half of the question came out of Blaine's mouth.

"Is that...?" Tracey tried to say but was too tired to finish.

Both Sassafras twins fell into an immediate, deep sleep. Summer looked at the dozing twins and laughed. She then looked at her excited lab assistant and shrugged.

"Sorry," the happy-go-lucky scientist apologized with a grin. "I guess you'll just have to show the twins your surprise tomorrow."

He landed in one of the facility's hallways. Luckily, he had been able to duck into a room before anyone saw him. He now stood in that room and laughed at how good his luck had been. Apparently, the room was some sort of control room because there were computer keyboards and monitors everywhere. It actually looked like the control room he had at 1108 N. Pecan Street, down in his basement. On the many monitors, he could see what was happening all over Summer Beach's underground science complex.

It truly was an impressive place, with all the latest technology. He knew for a fact the twins had only seen a small fraction of the huge underground facility. He wasn't sure why Summer hadn't shown them more of it. She was brilliant, but she could also be a bit forgetful—not nearly as absent-minded as his nemesis Cecil Sassafras, mind you, but forgetful nonetheless.

Thankfully, forgetfulness was not something that he struggled with, not at all. He was very good at remembering. His mind began to drift back, recalling once again the time in which Cecil had wronged him. He remembered how it felt and the pain of it all caused his rage to burn. He slammed his fist on the table, pulling himself out of the memory.

He muttered out loud to himself, "Yes, I remember that time and it is the very thing that drives my vengeance. I will pay you back, Cecil Sassafras!"

CHAPTER 15: INVASION OF THE ARCTIC GROUND SQUIRRELS

Haywire Hormones

Blaine and Tracey weren't sure how long they'd slept, but what they did know was that when they woke up, it was the next day and they felt invigorated, rested, and ready to go. They now found themselves sitting at the workstation, playing around with their new microscope apps again, having a conversation with the bright-eyed, frizzy-haired Summer Beach.

"Yes, so we started going to the same school together in the sixth grade," she was saying. "I was new to the town and the school. I was shy, and I was very nervous about starting classes at a brand new school."

"You were shy?" Blaine asked, not really believing that was possible.

"Oh, I was so shy," Summer confirmed. "And the first person that was nice to me was your uncle. He noticed that I was new and looking very nervous, so he invited me to sit at the lunch table with him. He looked at me with those dreamy blue eyes and started talking about science; my heart just melted."

"Cecil Sassafras!" Summer exclaimed, closing her eyes. "What a manly hunk!"

Both twins' faces scrunched up. They had both grown to love their uncle, but manly hunk? They didn't really think that described their uncle very accurately.

"How can you say Uncle Cecil is a manly hunk?" Tracey questioned, laughing.

"Oh, he's a total hunk," the scientist repeated, completely

convinced. "You know, other than being a brilliant scientist, he is also a black belt in Brazilian jujitsu?"

"What? Brazilian jujitsu?" the twins asked, shocked. "There's no way!"

"It's true," Summer confirmed. "When we were in seventh grade, a group of bullies decided they were going to pick on Cecil every Friday as he was walking home from school."

"I despise bullies," Blaine interjected, shaking his head.

"Me, too," Summer agreed. "There are so many bullies in junior high and high school, probably because during adolescence everyone's hormones seem to be going haywire. Anyways, I know those bullies were hard on your uncle. For the whole fall semester, Cecil tried and outrun them if he could. By winter break, he was fed up with the bullies and was tired of running. He did some research, and he came to the conclusion that jujitsu was the most scientifically and mathematically sound of all the martial arts. So, he began training in that discipline."

Summer continued, "By the time the first Friday of the spring semester rolled around, Cecil was ready for those bullies. As soon as they saw he had been training in jujitsu, they were the ones who were running away. Cecil continued to work his way through the belts until he received his black belt. He only used his jujitsu skills if he needed to defend himself, but it turns out he preferred fight over flight."

"Fight over flight?" Tracey asked. "What does that mean?"

"These are the body's two automatic responses to stress: fight or flight," Summer explained. "First, the body perceives a source of danger. Then it sends a signal to the adrenal gland, which in turn releases adrenaline causing our heart rate and breathing to increase so that our body will be ready to respond to the danger by fighting or running away."

"Fight or flight, huh?" Blaine mumbled, thinking about

what Summer had just said. "I guess we Sassafrases are more fighty than flighty, but it sure is hard to believe that Uncle Cecil has a black belt in Brazilian jujitsu."

"Well, that hunk of a man does." The female scientist smiled.

Summer hopped up from her seat and dance-walked over to one of the floor-to-ceiling translucent data screens. "Now I will help you two learn a little more about hormones and the Endocrine System."

Summer swept her fingers around the screen until the appropriate data page was visible and then began reading. "Hormones are chemical messengers produced and released by the glands of the endocrine system. Hormones control most of the body's processes, such as regulating mood, metabolism, and... oh, fiddlesticks!" she abruptly stopped.

"I totally forgot! I was supposed to call Ulysses S. Grant when the two of you woke up, so he could show you his surprise!"

The twins remotely remembered something about a surprise from Ulysses, but they had been so groggy with sonic lag the night before that the memory was a little hazy. Summer ran over to the wall and pushed an intercom button.

"Ulysses!" she shouted into the intercom. "Blaine and Tracey are up and at 'em and ready for your surprise!"

Summer raced back over to the table, looked at the twins, who were still looking at the slides before them, and exclaimed, "You two are going to love this!"

Moments later, the big sliding door began to move upward and open. Blaine and Tracey heard something before they saw anything. The sound was like hundreds of miniature feet hitting the ground. Then, the smiling Ulysses S. Grant appeared, followed by just that: hundreds of miniature feet that belonged to a whole crowd of robot squirrels!

THE SASSAFRAS SCIENCE ADVENTURES

They scurried into the room with movement that was robotic yet still very life-like. They began playfully running around the lab, climbing on the tables and chairs, and running up the display tubes. They even hopped up onto Blaine and Tracey. Summer giggled and clapped at the spectacle.

"It's a great surprise, isn't it!" she exclaimed.

The Sassafras twins nodded in wonder.

"Ulysses has been working on these little guys for a while," Summer shared. "It's an idea he came up with one day when he was really missing all of his squirrel buddies from out in the wild. He really loves working down here in the lab, but he is an arctic ground squirrel, after all. So, he decided to make some robotic squirrel companions, but they are more than that. They have the ability to climb up almost any surface and burrow through almost any material. Additionally, they are equipped with cameras and internal specimen storage units. All of this gives them the ability to go almost anywhere to take pictures and store small scientific specimens of things on which Ulysses S. Grant and I are doing research. It's simply fantastic!"

"It is fantastic," the twins agreed.

Tracey reached down and picked up one of the robot squirrels. It was just a little bigger than her hand and was very intricately designed with hundreds of small metal parts. It waved its big tail and looked at her with its blinking green eyes.

"How cute!" the Sassafras girl expressed.

Blaine picked up one too as he exclaimed, "These things are so cool!"

He set it down and watched it run off across the floor.

"And fast," he added.

The robot squirrels continued to playfully dart around in the lab while Ulysses watched proudly from a tabletop. The twins still really didn't know what to think about Ulysses S. Grant, or President Lincoln, for that matter. They knew they liked the furry animals, but were they both really inventors and lab assistants? It just didn't seem possible. Then again, neither did invisible zip lines that could take you around the planet at the speed of light. Their summer continued to be full of impossible surprises.

While the twins had been sleeping comfortably in their pods, he'd spent the majority of the night trying to come up with the best sabotage plan. Now he thought he had it.

Originally, he'd thought if he could just find some kind of power source or electrical box, he could shut down the electricity to the complex, but there was nothing like that in this room. What this room did have besides the monitors and keyboards was some sort of strange remote control. He had found it lying next to one of the keyboards. It was covered in buttons and levers, most of which he hadn't touched yet, but he'd played around with it a bit and it looked like it controlled the small army of robot squirrels.

He was watching right now on one of the monitors as those twins laughed and ran and played around with the robotic squirrels, but what bothered him wasn't the twins' happiness. Rather, it was the fact that Summer Beach had just found the hormone information on one of her fancy data screens, and she was about to share that information with those twins. He must put his sabotage plan into action right now. He would use the remote to control those squirrels and torment those twins!

"Now, where were we?" Summer Beach asked, looking at the hormone information on the data screen. "Oh, yes, the processes that hormones control."

Blaine and Tracey walked over closer to the screen. Tracey was still holding one of the cute green-eyed robotic squirrels in her hand. She was having a little trouble paying attention to the SCIDAT data because she was enamored by the cute little guy she was holding. Its life-like movements and many complex parts truly were amazing.

Suddenly, something very strange happened. The cute little robot squirrel's eyes turned from green to red, and he reached down and bit Tracey on the hand.

"Ouch!" Tracey yelped as she dropped the robot on the ground. "The robot squirrel just bit me."

"What?" Summer asked, alarmed.

"He bit me!" Tracey repeated, pointing at the culprit squirrel as he scurried off to join the rest of the robots.

The three watched the squirrels as they started acting

strangely, flipping over, bumping walls, and using their metal teeth to bite each other. Then, slowly but surely, they huddled together as a group. They stayed in their huddle for only a few seconds, and then they turned as a group toward the three humans, all of them now with glaring red robot eyes. Ulysses S. Grant stood on the table top looking worried and concerned. Then, to Blaine and Tracey's dismay, the robot squirrels started running toward them.

"Fight or flight?" Summer shouted in question form.

"Flight!" the twins screamed together.

Summer reached out and smacked a button with her hand and the big egg-shaped door began to slide open. The scientist and the twins dove under the door before it was fully open and began sprinting down the long corridor.

"I know robots don't have hormones," Summer sputtered as they ran. "But it's almost like those squirrels do because of the way they suddenly got out of whack!"

The red-eyed robots began streaming out of the lab and into the corridor behind them, chasing at a high rate of speed. Blaine looked back and saw that Ulysses had jumped down off the table and was trying to block the robots from chasing anyone, but there were just too many robots. They easily ran over the worried arctic squirrel.

The twins recognized this corridor as the one they'd walked through the week before to go to the heliquickter, Summer's ultra-fast helicopter. They wondered if that was the woman's plan—to escape in the heliquickter. That question was answered quickly, though, when instead of running the length of the corridor to the heliquickter, Summer pushed another button on the wall. It opened up to a long hallway that led off to the right. The twins quickly followed the scientist into this hallway as their sprint continued.

"Hormones control processes in the body that happen over a long period of time, such as growth," Summer shared as they ran.

"They're released into the bloodstream where they travel around until they reach a specific type of cell. Once there, the hormones transfer information and chemical instructions to the target cells."

Summer pushed another button. The twins followed her, bolting through an open doorway to their left. They could hear metallic feet still chasing closely behind them.

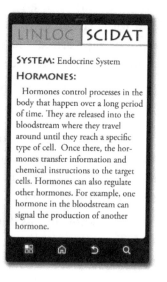

SYSTEM: Endocrine System
HORMONES:

Hormones control processes in the body that happen over a long period of time. They are released into the bloodstream where they travel around until they reach a specific type of cell. Once there, the hormones transfer information and chemical instructions to the target cells. Hormones can also regulate other hormones. For example, one hormone in the bloodstream can signal the production of another hormone.

"Hormones can also regulate other hormones," Summer continued. "For example, one hormone in the bloodstream can signal the production of another hormone."

The twins weren't sure how Summer could concentrate on science at a time like this. The snipping, chasing robot squirrels seemed to be getting closer and closer.

"We have over thirty hormones in our body that regulate many things," Summer informed. "Like, when we are hungry, when we are sleepy, how we break down food, how we handle stress, how we grow, the temperature of our body, and so much more." The scientist skidded to a stop and grabbed the rungs of a metal ladder built into the wall.

"That's about everything that was included in the hormone data. If we ever make it back to the lab, I will upload it to your phones."

Summer then climbed the short ladder and opened up a square metal door in the ceiling. She climbed through the open door then reached down a hand to help the twins up. "Maybe we can throw off these squirrels if we go up a level," she offered.

The Sassafrases used the rungs and Summer's hand to quickly

exit the hallway they were in. They peeked down through the open door and watched as the robots ran by underneath them.

"Yes! It worked!" Blaine exclaimed in a loud whisper, but he had spoken too soon. The three heard the footsteps of the robots stop. The squirrels doubled back and looked up the short metal ladder at them. Immediately, the robots started climbing the ladder in pursuit.

"Rutherford B. Hayes!" Blaine scream-shouted.

They were now in a small room with four doors. Summer pushed one open, and the three plunged through the doorway, landing in some sort of utility room. There were all kinds of tools, extra supplies, and mismatching metal parts.

"This may be the very room where Ulysses S. Grant invented these robot suckers," Tracey thought to herself as she raced through the room with Blaine and Summer. On the other side of the room, Summer reached up and grabbed a hammer that was on the wall.

"Mind your arms and legs," she warned as she pulled the hammer's handle.

"Wha—?" the twins half-asked before they realized they'd fallen through the floor and were now sliding down a well-lit tube slide.

Unfortunately, yet again, the robot squirrels were not thwarted by the three fleeing humans' change of direction. They jumped down the slide one after the other, giving chase.

The bright lights that were embedded in the walls seemed to make noise as the three zipped down past them—zoomp, zoomp, zoomp. They crashed through a swinging door at the end of the tube-slide and landed in a stumbling run inside of another long hallway. The robots came shooting out of the slide right on their heels.

Summer pressed on, straight ahead, leading the way.

THE SASSAFRAS SCIENCE ADVENTURES

She turned back to the twins as she ran and shouted, "When I say lunge and jump, lunge and jump!"

The twins didn't really understand, but they nodded nonetheless. After a few more big strides, they heard Summer doing a countdown.

"Three… two… one… lunge and jump!"

The scientist took a big leap forward and planted both of her feet firmly on the surface of the hallway. She then bounced up and burst through a circular opening in the ceiling. The twins didn't know how this was possible, but they mimicked her exact movement, a lunge and jump, and they too found themselves bouncing up through the hole in the ceiling.

"Was that a trampoline floor?" Blaine asked when he landed safely joining Summer and his sister.

"Yes, it was," Summer confirmed, smiling, but still looking a bit worried.

The three looked down through the circle that was now in the floor. The robot squirrels had reached the trampoline floor, but they'd momentarily stopped.

"They won't be able to bounce on the trampoline to get up here, but they will be able to climb the walls. C'mon Sassafrases, let's keep going!" Summer ordered.

The twins followed Summer as she began racing up a spiral staircase. Around and around they went until they reached the next level. The robot squirrels had already climbed the walls and had already made their way through the circular opening. They were now spiraling up the staircase behind them like an upside-down tornado.

"Those things are unstoppable," thought the twins as they ran on. The room they were in now looked like some kind of huge cockpit. They quickly ran out of that room, through an egg-shaped doorway, into the next room, which had a giant telescope right in the middle of it.

"More slides!" Summer shouted as they ran toward a wall with four huge circular openings on it. "These four slides go to four different places. If we all jump in the same one, then maybe the squirrels will pick the wrong one and lose us all."

Summer dove head first into the farthest circle to the left. The twins followed her—zoomp, zoomp, zoomp—the three went, sliding down another lit tube slide. All three landed in somersaulting crashes in a…deli?

"Is this a deli?" Tracey asked the obvious question.

"Sure is," Summer grinned. "I love sandwiches."

The three sat on the floor quietly for a second, intently listening to see if the robot squirrels had picked the correct slide. For a moment, it was all quiet, leading the three to believe that maybe they had lost their robotic pursuers, but then they heard it: the sound of metal coming down the slide toward them.

"Follow me!" Summer shouted, jumping up. "I have an idea."

The scientist ran behind a counter that was stocked full with supplies to make an endless variety of sandwiches and grabbed the handle of a big stainless steel door.

"Just like many delis," Summer shared, "this deli is equipped with a walk-in freezer. We can shut ourselves in here away from the robots."

Though the twins had already been locked in a freezer once since studying anatomy, they saw no better choice. They were tired of being chased.

"Let's do it," Tracey agreed.

The robot squirrels began appearing in a pile at the bottom of the slide. They immediately spotted the three with their red eyes and darted toward them. Summer yanked open the huge metal door and hurriedly helped the twins inside before jumping in. She slammed the door shut, just before the robots got to her.

Fight or Flight

He slammed his fist down in anger on the desk next to a keyboard. He had almost gotten them again! Now they were secured behind some kind of huge metal door.

"Arrgh!" he exclaimed, frustrated.

He had figured out how the remote control worked rather quickly and had learned how to maneuver and control the robot squirrels. They were fast little robots but not quite fast enough. He'd chased those twins and that scientist through a big portion of the underground science facility. He'd managed to stay on their heels, but he just couldn't quite get those little squirrels to catch them.

The monitors in this room had been helpful. There were so many, in fact, that as soon as the image of the twins and the scientist ran off of one monitor, they would appear on another one. So chasing them with the robots had been fairly easy, but he still hadn't managed to catch them.

He looked at one of the monitors right now and saw that the three were stuck in a dead end room behind the big metal door. It looked like maybe it was a freezer. This was his best chance to catch them—if he could just figure out a way to get through that door. There were still a few buttons and levers on the remote control that he hadn't tried. He mused out loud as he reached his fingers out to pull a new lever.

"I wonder how sharp the teeth on these robots are."

"What is that sound?" Tracey asked, shivering and alarmed.

Summer looked at the freezer door. "Do you remember how I told you two that the robot squirrels were designed with the ability to burrow through almost any kind of material?"

The Sassafrases nodded, remembering.

"Well, I think they're doing that just now," Summer informed. "I think the robot squirrels are burrowing through the freezer door."

The twins were shocked. They had thought they were safe. It just didn't seem possible that small little robot squirrels could actually dig through a thick insulated freezer door, but even now, they could hear the sound of grinding metal as the robots worked together like a rabid pack, burrowing in from the other side. Summer looked around the icy room.

"Well, we couldn't have gotten stuck in a better place," she smiled.

"What?" the twins thought. "How is being stuck in a walk-in freezer good in any way? Especially with crazy robot squirrels burrowing in to get you." Neither Sassafras voiced their doubtful thoughts because they figured that Summer was probably about to explain herself.

"Just like most deli freezers, this one is stocked full with all kinds of delicious sandwich meats, but unlike most deli freezers, this freezer also has two special doors on the back wall." The scientist's eyes twinkled as she spoke.

The twins looked at the back wall of the freezer. At first, all they could see was a metal wall covered in frost, but then, looking more intently, they could make out the seams of two small doors, one on top of the other.

"What makes those doors so special?" asked Blaine.

"The bottom door leads to an escape tunnel," Summer explained. "The top door leads to something completely different. Behind that door there is a rack of ice guns."

"Ice guns?" the twins asked.

"Yep, ice guns," Summer confirmed. "Yet another one of Ulysses S. Grant's super cool inventions. We have used them for all kinds of fun things."

The sound of grinding metal seemed to be getting louder. The twins were sure the robot squirrels would bust through the freezer door at any moment. They were ready to open up the bottom door and race into the escape tunnel. On the other hand, Summer didn't seem concerned at all as she continued talk about the ice guns. "The two main things we do with the ice guns are freeze any scientific specimens that we need to preserve, and we freeze the sandwich meats," the bubbly scientist giggled.

Blaine was growing impatient. He stepped toward the back wall ready to yank open the escape tunnel door.

"I think I know another way that we could use these ice guns right now," Summer continued. "I think we can use them to freeze the robot squirrels."

Blaine stopped thinking about the escape door for a second, and he looked back toward the scientist.

"We can freeze them?" Blaine questioned.

"I think so," Summer responded. "The ice guns usually freeze whatever we are aiming at instantly."

"Instantly?" Tracey asked.

"Instantly," Summer confirmed. "So my question to you, Blaine and Tracey, is this: fight or flight?"

He didn't know exactly how thick the door was, but he was certain that the robots had almost successfully burrowed through it. He couldn't hear what the three were saying, but he could still see them on the monitor. Summer looked fine, but those twins looked scared and nervous. His heart was beating fast. He was finally about to catch those Sassafras twins.

Blaine and Tracey looked at each other. They were scared, but they knew what their answer should be. The Sassafras twins shook away their shivers and then faced Summer with their answer.

"Fight!" they both shouted.

Summer smiled and jumped to the freezer's back wall where she yanked open the upper door. She reached in and grabbed two ice guns off of a rack and gave them to the twins. The twins held the ice guns in their hands and looked at each other with raised eyebrows. They had expected these things to look a little scarier than this. These guns looked just like cheap brightly colored water guns that you could buy at a discount store.

"You're sure these little things can freeze a robot squirrel instantly?" Tracey asked the scientist doubtfully.

"Pretty sure," Summer responded as she grabbed an ice gun for herself.

The three faced the freezer door and all aimed their ice guns straight ahead. The sound of grinding and ripping and shredding metal continued until, with a small spark, they all saw them: a pair of red eyes.

"Yes!" he shouted in elation. "Those little metal pieces of junk made it through!"

He watched on the monitor as the first robot squirrel punched his head through the freezer door. The scientist and those twins had grabbed some kind of plastic water gun-looking things out of a door in the wall that he hadn't seen. It was laughable that they thought they could really defend themselves with those things against an army of metallic robot squirrels.

He pushed on the levers of the remote control, directing the squirrels forward. He was so busy staring at the monitor and directing the attack that he didn't notice when the door to the room he was in opened slowly behind him. Before he knew it, he was being attacked… by a squirrel… an arctic ground squirrel.

The first robot squirrel burst through the door and jumped into the air straight toward Tracey. The Sassafras girl froze, but thankfully Summer didn't. She aimed her ice gun right at the little flying machine and pulled the trigger. The metal robot landed with a clunk on the freezer floor, encased in a solid block of ice.

"Whoa," Blaine whistled in amazement. "These ice guns really work!"

"Let's use them," Tracey snapped back to life.

The robots began to pour into the walk-in freezer through the hole in the door. As they did they were getting zapped with ice guns, one by one.

One would think that fighting off the attack of an arctic ground squirrel would be easy, but it wasn't. Especially when that arctic ground squirrel's name was Ulysses S. Grant. He was frantically swinging, punching, stomping, and kicking at the pesky vermin, but Ulysses was just too quick and elusive, and the squirrel's one-two punches were surprisingly painful. One moment Ulysses would be on the table in front of him; then he would be on his leg, then his back, and now what was this? Was the ground squirrel doing some sort of jujitsu move on him?

Ulysses S. Grant grabbed his arm and twisted it to a weird angle, causing him to lose his grip on the remote control. The device slipped from his fingers and in a flash Ulysses was right there to pick it up, but before the ground squirrel could get away, he got a hand back on the remote. Now, the wrestling match began between the eyebrow-less man and the squirrel over the ever so important remote control.

Summer Beach, was jumping, laughing, dancing, flipping, and shooting like she was playing a game at an arcade. The robot squirrels were streaming into the freezer at a steady flow until, suddenly, they began to act erratically.

"What's happening?" Tracey asked.

"I don't know," Summer answered, while shooting a behind the back shot with her ice gun. "But just keep aiming, shooting, and freezing!"

Now, instead of coming straight toward the three, the robots began to abruptly stop and start, move from side to side, and even run face-first into the racks of frozen sandwich meats. They would look like they were attacking, then look like they were backing off, attacking, and backing off again. The crazy machines sure did seem to have dual personalities.

The three continued to ice the robots away, causing the stack of squirrel-filled ice blocks to mount higher and higher. The Sassafrases' fear began to slowly melt away. This was actually quite fun; shooting robots with ice guns. Then the twins both noticed something at the same time.

"Look!" they pointed out to Summer. "The squirrel's eyes have changed back to green!"

"Well I'll be!" the scientist exclaimed as she holstered her ice gun. "They sure have."

It had been an epic battle, but in the end, it had been the arctic ground squirrel that had risen victorious. He now found himself lying on the floor, scraped, bruised, and defeated. Ulysses S. Grant had secured the remote control and had scurried out of the room. He slowly sat up and looked at the monitors. He saw that the three were now doing a victory dance in the freezer. He had failed again; surely this was the longest losing streak by a villain ever. He reached over and grabbed his harness and carabiner. It was time to zip back to 1108 North Pecan Street and hit the drawing board.

The twins looked at the remote control lying on the table. They were with Summer Beach and Ulysses S. Grant, and were safely back in the lab.

"So this thing can control the squirrels?" Blaine asked.

"And it went haywire?" added Tracey.

"I suppose so," answered Summer. "Each robot has an individual on/off switch, but Ulysses added this remote to their control system. His plan was to slowly but surely add to their skills, ability, and movement. That's why there are so many buttons and levers on this thing. I'm not sure exactly how it happened; maybe the frequencies got crossed or maybe there was some sort of other malfunction, but whatever it was, it is fixed now and our robot squirrels are back to their normal friendly selves. As soon as they thaw out, we will test them for sure, but for now everything seems to be fine."

The Sassafras twins sighed. They were glad that the chase was over and they were happy for Ulysses and his robot squirrels. They hoped Summer was right about the robots really being back to normal. One thing about the crazy chase though, was that they had learned a lot about their fight or flight instincts.

"Well, I guess we did a little 'fighting' and a little 'flighting' this morning." Blaine laughed.

"We sure did!' Summer grinned. "It has been a pretty crazy morning, but at least it affords us the opportunity to learn something."

The scientist refreshed the data page with the hormone information and pushed the upload tab. Blaine and Tracey immediately felt their smartphones vibrate with SCIDAT data.

"We finished covering all the hormone data on the run this morning," Summer reminded. "Now let me tell you about the glands; one of which is the adrenal gland, which plays a big part in your fight or flight instincts."

Summer flipped over to the gland information page on the screen, but before she started reading it she did a joyful jump-kick clap-dance combination move as she exclaimed, "Oh what adventure we had this morning! Adventure and science all mixed together! Does life get any better than this? I think not!"

Blaine and Tracey suddenly both found themselves jumping, clapping, and dancing. Evidently Summer Beach had rubbed off on them.

They eventually calmed down and the scientist scrolled to the lower-middle of the gland data, where she found information on the adrenal gland.

"The adrenal gland," she started, "is actually a pair of glands that sit on top of the kidneys. Each one produces hormones that help regulate the body's metabolism and response to stress, which includes the fight and flight response. So that's the adrenal gland."

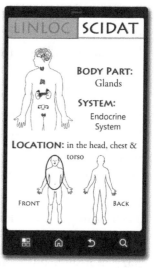

Summer scrolled back up to the top of the page. "There are eight more glands that we will learn about here. So altogether, it makes nine main glands that produce and release hormones into the body. All of which are found in the head or the torso of the body."

Summer began reading from the top of the page: "The pituitary gland is about the size of a pea and is attached to the brain. It releases six different hormones and is known as the master gland because it controls the growth of the body and the production of other hormones. The pineal gland is in the middle of the brain. It releases melatonin, a hormone responsible for regulating our sleep patterns. The hypothalamus is a gland located at the center base of the brain. It takes part in regulating appetite, metabolism, and body temperature."

Summer pointed to each of the glands on the screen as she

shared about them. "The thyroid gland is a butterfly shaped gland that is located in the neck. It makes two hormones which regulate metabolism and growth. The parathyroid gland is a pair of glands that sit right on top of the thyroid gland. They release hormones that regulate the amount of calcium in the blood. The thymus gland is located under the sternum in the upper torso. It shrinks as we approach adulthood, but during the early years it produces a hormone that ensures the normal development of T-cells."

The twins could see that they had now reached information on the adrenal gland. Summer skipped over that, since she had already covered it and read about the next glands.

"The testes and ovaries are glands that release hormones that make a human being look like either a man or a woman. Males have testes and females have ovaries. They are located in the lower half of the torso. And finally, we have the pancreas, which sits behind the stomach about mid-torso. It produces and releases hormones that regulate blood sugar levels, as well as several digestive enzymes."

Summer finished reading the data and scrolled down to the upload button.

"Okay! That's that!" she exclaimed as she clicked the tab

LINLOC   SCIDAT

LOCATION: Dubai, UAE
CONTACT: Sylvester Hibbel
LATITUDE: LONGITUDE:
25° 15'N   58° 18'E

DUBAI

INFORMATION NEEDED ON:
Skin, Sweat, Hair,
and Fingernails

The twins received the data and then finished completing the leg's requirements by adding all the appropriate pictures using either the archive app or their new microscope app. They sent the SCIDAT data into Uncle Cecil and then opened up their LINLOC apps to take a peek at where they would be going next.

Summer giggled with delight throughout the whole process, while Ulysses S. Grant just sat and watched the retrieved ice blocks slowly melt away from his beloved robot squirrels.

"We are going to a place called 'Dubai' in the United Arab Emirates!" Tracey shared enthusiastically, as she read the information. "Longitude 55° 18' E, Latitude 25° 15' N."

"Where we will be gathering data on the Integumentary System." Blaine added. "More specifically; skin, sweat, hair, and fingernails. All with the help of a local expert named Sylvester Hibbel."

"Sassafrases! Sassafrases! Sassafrases can! Learning science while zipping across the land!" Summer shouted, clapping like a lab coat wearing cheerleader.

"That's a little cheer I made up for you two." The female scientist smiled like a proud parent.

Summer sure was crazy, but both Blaine and Tracey really liked her, and they were actually really going to miss her.

'Will we see you again?" Tracey asked as she and Blaine both slipped on their harnesses.

"Only time will tell," Summer teased. "You two know how the zip lines work. You don't know where you're going next until you successfully complete your SCIDAT data at each location, but as you saw today, there is a lot more to my science facility than just this lab. There are many doors we left unopened, and there is science here that we have yet to uncover, so you never know when you might see my name pop back up on your LINLOC screen."

Just the possibility of seeing the friendly scientist again made the twins happy. They were enjoying all the new and bold adventure, but a little familiarity wasn't bad either.

The twins calibrated their carabiners and let them snap shut on the next invisible zip line. As they hung in the air and waited for zip to happen, they said their goodbyes to Summer and Ulysses. Then, in a flash, they were off, with big smiles on their faces. They flew through bright white at the speed of light. After a few seconds of zipping they landed, but the landing took the smiles right off their faces, because once again they had landed in water. The last time they had done this, it had been in the South Atlantic, and they had nearly drowned.

"Oh no, not again!" the twins thought in fear. "How long will we have to hold our breaths this time?"

CHAPTER 16: THE WIND TOWER 100

Old Doc Hibbel's Hide-a-Chap Balm, Great for the Skin!

"Sweet Sarsaparilla!" the twins heard a man's voice say as their heads popped up out of the water. "If you two were thirsty, you should have come to my wagon, not the drinking troughs!"

The Sassafrases could now breathe, but their sight was still gone—as was their strength.

"Well, I'll be Roy Rogers' uncle!" the voice exclaimed. "You two young 'uns don't look very well, but, not to worry, I've got just the thing for you."

Both Blaine and Tracey felt themselves being pulled out of the shallow troughs of water that they had landed in. Their limp bodies were set on the ground.

"Hold on, you two. Doc Hibbel is on the case!"

The twins heard the man take a few steps off to the side, and then the sound of him rummaging through something like glass bottles could be heard. Their sight and strength was slowly beginning to return, but neither had returned fully before the man crouched down beside Blaine, propped him up, and poured something into his mouth.

As soon as the liquid reached Blaine's throat, he shot up to his feet with a shout. His vision and strength were now fully restored. "What was that stuff?" Blaine coughed. "It tasted like motor oil with some sugar mixed in!"

The man moved over to Tracey and poured the same liquid into her mouth. Immediately, Tracey shot to her feet with the same look of disgust on her face that Blaine had. The man they could now see in front of them had no hint of disgust on his face. Instead, he was smiling from ear to ear.

"That was Old Doc Hibbel's Fixer Elixir," he grinned, holding up a small glass bottle for the twins to see.

The Sassafrases both smacked their tongues with open mouths. It still felt like their whole mouths and throats were coated with the stuff. "This must be our local expert, Sylvester Hibbel," thought the twins. He looked like he'd just stepped out of a western movie. He had a thick salt-and-pepper-colored mustache and was wearing a wide-brimmed cowboy hat, a cowboy-style bowtie, black dusty boots, and spurs, chaps and a sleeveless vest.

"This elixir has been known to cure a wide range of ailments," the doc continued, "such as loss of vision, loss of strength, loss of memory, loss of wallet, back aches, tooth aches, toe aches, and heartaches. It can put hair in your step and hop on your head. It has been known to ward off savages, stop tornadoes, still earthquakes, and leave a real nice shine on shoes. Old Doc Hibbel's Fixer Elixir!"

Blaine and Tracey had no idea how to respond to their new local expert's sales pitch. They weren't sure if the elixir had caused them to shoot out of their stupor or if they were already equalized before the elixir hit their throats. All they knew was that it tasted terrible.

The twins took a look around at their surroundings. They thought they were supposed to have zipped to the United Arab Emirates in the Middle East, but this place looked more like they had zipped back in time to the Wild West days of the United States. There were wooden buildings on either side of them, making them feel like they were on the small dusty main street of an old western town. There was a post office, jailhouse, Goods and Mercantile, saloon, bank, general store, restaurant and stable. There were quite a few horses tethered to wooden posts close to the drinking troughs that they had landed in after their zipping. And here, right in front of them, was the western peddler, Sylvester Hibbel, standing next to his horse-drawn covered wagon that had a sign with neatly painted, western-style lettering that read, "Old Doc Hibbel's Western Goods."

Sylvester took off his hat and held out his hand to officially greet the twins. "My mother named me Sylvester Robert Hibbel. My satisfied clients call me Doc. My wife calls me selfish or grumpy, but you two may call me friend."

Both twins reached out and shook Sylvester's hand. "We are the Sassafras twins," Blaine offered. "Blaine and Tracey Sassafras."

"Well, very nice to meet you," the peddler said. "Have you two come here for the race?"

"What race?" Tracey questioned.

"What race!" the man said like it was preposterous that the girl didn't know what he was talking about. "Why the 'Wind Tower 100' of course!"

"The Wind Tower 100?" Tracey asked, still obviously in the dark.

"I don't know why you two were swimming in the drinking troughs or where you come from, but around these parts, The Wind Tower 100 is a pretty big deal."

"Where exactly are 'these parts'?" Blaine asked.

"The United Arab Emirates," Hibbel stated, "but to be more exact, Dubai."

"This is Dubai?" Tracey asked.

"Well, this isn't Dubai," Sylvester said, pointing down to the ground. "This is a fabricated town, built to look like the USA's Old West."

Sylvester then strode to the corner of one of the wooden buildings and pointed to something in the distance.

"But that, is Dubai," he announced.

The twins followed the man to the corner of the building and peeked around to see where he was pointing. The Sassafrases couldn't believe they hadn't seen it before now; because right in front of them stood the desert metropolis city of Dubai. They now saw that the fake western town was actually built in the parking lot of a huge mall.

"We're right at the edge of town," Hibbel explained, "with the city of Dubai on one side and the vast Arabian desert on the other. This is where the race starts tomorrow, right here in the parking lot of Wind Tower mall."

"What's so special about tomorrow's race?" Blaine asked.

"It's got quite a large winner's purse," Sylvester answered, "but there is much more to it than that. This race is so much bigger than something you can win. Just being able to compete in it and be a part of the race itself is the real prize."

Sylvester Hibbel smiled as he thought about it. "The Wind Tower 100 is the greatest distance horse race left in the world. There are several of these kinds of races around. America has a few and there are some in Europe and Asia as well. The longest current one happens annually in Mongolia. It's a one thousand kilometer race, which is close to six hundred forty miles, but it's completed over the course of eight to ten days, and the riders use more than one horse. That's what makes the Wind Tower 100 the most difficult distance race in the world. It's a trepid one hundred mile race across the unforgiving Arabian Desert. The race must be completed in one day using only one horse. It starts at the modern Wind Tower Mall, here, and ends one hundred miles away at an ancient wind tower in a city called Hatta."

"What is a wind tower?" asked Blaine.

Sylvester pointed up to the highest point of the mall.

"You see that ornate tower that looks like it has pipes or sticks sticking out of it?"

Blaine nodded.

"That is a wind tower," the doc informed. "It is designed to catch the cooler air from up higher and funnel it down into a building. Any relief at all that you can get from the scorching desert heat is appreciated here in the UAE."

"Are you one of the racers?" Tracey asked.

Sylvester chuckled at the question. "No, ma'am, I'm not—at least not anymore. My last distance race happened many moons ago when I was a much younger man. I wasn't any better at it then than I would be now. No, I'm merely a traveling salesman who had the good fortune to meet Sheikh Rehan a while back. The man is a billionaire, an oil baron, a real estate mogul, and a huge fan of distance horse racing. He founded the Rehan Equestrian Club here in Dubai. He also built the Wind Tower Mall, which you see standing in front of you. The Wind Tower 100 is put on and sponsored every year by the Sheikh, his mall, and his equestrian club.

"The Sheikh, like many of the Arab riders and race enthusiasts, is a big fan of Old West American cowboy culture. I am an expert on the subject, as I happen to be a trustee for the Cowboy Hall of Fame in Oklahoma City, Oklahoma. So, Sheikh Rehan flies me, my horses, and my wagon out here to Dubai every year to sell my old-fashioned western products. This year, to my surprise, he even had this old western town built to serve as tomorrow's starting line."

"This is all fascinating," Tracey mused.

"Yeah, it really is cool," Blaine agreed.

"Speak of the Sheikh," Sylvester said. "Here he comes now, and it looks like he's got some friends with him."

The twins looked and saw that a small crowd of people were exiting the mall and walking toward them. Most of them were Arab men, wearing their traditional head scarves and robes, but there were some women and some westerners mixed in as well. The man walking in front was wearing dark sunglasses and had a neatly trimmed black beard. He came right up to Sylvester with a smile, reached out, grabbed Hibbel's outstretched hand, and pulled the peddler close to touch his cheek to Sylvester's cheek on the right side, then the left. The twins thought that was a little strange, but hey assumed that was just the way people greeted each other here.

"Doc Hibbel! My dear friend!" the Sheikh boomed. "It is so good to have you here in Dubai again this year for the Wind Tower 100! How do you like the Wild West town I had built for you?"

"It's perfectly splendid," Sylvester smiled.

Sheikh Rehan then glanced back at the group of people behind him. "Doc, I brought some of this year's best riders out to meet you. Surely this year's winner will come from one of these jockeys behind me."

Some of the riders laughed at this statement, but others looked like they took the statement as a challenge.

"I know many of the riders want to see what kind of western wares you have to offer," Sheikh Rehan said. "Plus, all the horses will be kept out here in the stable tonight, so the jockeys wanted to see where their horses will sleep."

Hibbel nodded once to the Sheikh and then bounded like a much younger man than he was over to his covered wagon. He unlatched the back gate and immediately, a set of stairs folded down instantly transforming the back of the wagon into a sales booth. They tiny store had all sorts of small western goods like hats, spurs, bullwhips, belts, belt buckles, bandanas, jerky, a wide assortment of elixirs, and much more.

Some of the excited jockeys hurried up into the wagon, while others huddled around Old Doc Hibbel as he touted his merchandise. It was fun for the twins to see traditionally dressed Arab men trying on cowboy hats, belts, and belt buckles right over their clothing. The Sassafrases got to meet several of the racers who were kind of like celebrities here in the UAE because of the prestige of horse racing.

They also got to meet and sit on some of the horses. One particular jockey, a light blond-headed westerner from the Netherlands named Arnie Derbinhoogan, even let the twins ride up and down the Old Western Main Street a couple of times on his horse. It was a horse by the name of Horsinhoogan.

The twins had overheard some of the other jockeys talking about Arnie and his horse. Evidently, the Dutchman was a favorite to win the Wind Tower 100, even though it was his first time to race in it. He was a champion in the European distance riding circuits, so therefore they saw him as a threat to take the winner's purse.

"I see you have quite a fair complexion," Sylvester noticed, as Arnie stood and looked over his western merchandise.

"I sure do," the Dutchman confirmed. "I have been a little bit worried about how my light skin is going to do under the hot Arabian sun."

"Well, have no fear, pale rider," Sylvester assured. "I have just the thing for you: Old Doc Hibbel's Hide-a-Chap Western

Balm."

The peddler pulled a flat circular tin off of the shelf, opened it up, and handed it to Arnie.

"If your lips, your hands, or any other part of your hide is chapped, and you want to hide that chap, then just generously apply Old Doc Hibbel's Hide-a-Chap Western balm to the affected area, but that is not all," Sylvester exclaimed. "This balm is also effective in protecting your hide from sunburn."

"Did you know that the color of your skin is due to the amount of a pigment called melanin that is found in the lower layer of the epidermis? The more pigment there is, the darker your skin will be. This pigment also helps to protect the body against the sun's harmful ultraviolet rays. Just one look at you, my good man, and one can tell that your amount of pigmentation is low. So, Old Doc Hibbel's Hide-a-Chap Western Balm should surely be a help to you."

Arnie took some balm out of the tin with his fingers and rubbed some of it on his skin.

"The skin has three layers," Hibbel continued. "The epidermis, the dermis, and a layer of fat. The upper layer of epidermis consists of flat dead cells that interlock to form a barrier or protective covering. The cells in this layer are constantly worn away and replaced. Close up, this layer looks like a criss-cross pattern of lines with small holes or pores in it, but under a microscope it looks scaly and flaky. I'll tell you something that will really chap your hide: every year, nine pounds of these flakes are worn away from your skin!"

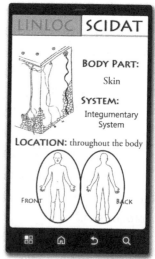

LINLOC **SCIDAT**

BODY PART:
Skin

SYSTEM:
Integumentary System

LOCATION: throughout the body

FRONT   BACK

"Nine   pounds,"   exclaimed

Blaine. "That is crazy!"

Tracey opened up the microscope app on her phone and used it to see for herself what Doc Hibbel was talking about. Sure enough, the close-up version of her hand that she could see on her screen looked scaly and flaky. She took a picture that she was sure she could use for SCIDAT later.

"The dermis is the lower, thicker layer of skin," Sylvester went on. "It contains the sweat glands, sebaceous glands and blood vessels. It also contains the sensory receptors that help the body to detect changes in temperature, touch, vibrations, pressure and pain. The layer of fat under the dermis helps to keep the body warm and absorb shock. Altogether, the skin is only a tenth of an inch thick, but it forms a barrier between the insides of the body and the outside world and its germs. The skin also protects the body from drying out and helps to regulate the body's temperature."

"So this balm will hide my chaps, protect me from getting sunburnt, and generally aid the health and well-being of my skin?" Arnie Derbinhoogan asked.

"That's right," Sylvester assured. "I give Old Doc Hibble's Hide-a-Chap Western Balm my own personal stamp of approval."

Arnie chuckled, knowing that he was being eloquently sold a product. "I'll take a tin," the Dutch jockey announced. "How much is it?"

"The first tin is on me." Sylvester tossed Arnie a new unused tin. "I know this is your first time to ride in the Wind Tower 100. So consider this a welcome gift. Good luck, son."

"Thank you very much, Doc," the rider replied gratefully.

The twins noticed as they watched Sylvester 'Doc' Hibbel conversing with the racers at his wagon that he gave away about as many products as he sold. He truly was a fan of Dubai's Wind Tower 100 race, and he was here more for the spectacle of the adventure rather than for making any money.

After talking with all the jockeys and showing them his miniature western town, Sheikh Rehan returned to the covered wagon with an invitation for his American friend.

"Doc, please come and stay at my recently completed five-star hotel tonight. The Desert Horizon is Dubai's newest and most luxurious hotel. It is just on the other side of the mall, and I have already made arrangements for you to stay in the Presidential Suite."

The twins could tell by the look on Sylvester's face that he hadn't been expecting this. "Well, Rehan, that is awfully nice of you and all, but I usually just sleep in my wagon."

"Oh, please, cowboy," the Sheikh insisted with a smile. "Don't decline an old friend's offer. Besides, I also wanted to talk to you about the possibility of actually driving your covered wagon out along the roadside checkpoints tomorrow."

Sylvester grinned at this part of the Sheikh's proposition.

"Okay, Sheikh," Hibbel agreed enthusiastically. "I would love to take the shop out tomorrow! I'll head over to the Desert Horizon just as soon as I finish up here."

Sheikh Rehan nodded and smiled, looking glad his friend had taken him up on his offer. He then climbed into the back of a Hum-V limousine that had pulled up in the parking lot and drove away. Sylvester looked at the twins with excitement in his eyes.

"I-yip-i-yo-ee-yay!" he shouted. "Looks like, for the first time, I will actually get to ride out with the racers!"

The Sassafrases smiled, happy for their local expert.

"Would you two want to come with me?" Hibbel asked the two. "I could use some help on the trail."

"Sure!" the twins exclaimed at the same time.

"Wow! The Wind Tower 100, the best distance race in the world, and we get to be a part of it," Sylvester marveled.

"What are the roadside checkpoints that the Sheikh was

talking about?" Tracey asked.

"As I mentioned before," Hibbel explained, "the race is an even one hundred miles, but there are several checkpoints along the way that the riders must check into as they progress through the race. Some of these checkpoints are out in the desert, and some are along a certain roadway. The road is a seventy-one mile route across the desert from Dubai to Hatta. Sheikh Rehan and other dignitaries and race fans watch the start of the race and then travel along the road in their vehicles to catch glimpses of the racers at the roadside checkpoints. Then, they drive ahead to Hatta and wait for the winner to come across the finish line. I'm hoping that the Sheikh wants me and my wagon at the roadside checkpoints to be able to offer refreshments and any needed western wares to the riders as they check-in."

"We'll be more than happy to ride out with you, and help you in any way we can," Blaine offered excitedly.

"Good," the salesman declared. "Then, if you two don't mind, since I will be staying at the Desert Horizon tonight, I'll ask you to sleep in the covered wagon to guard my horses and goods. I think everything should be fine, but this is the Arabian Desert after all. It always seems to have a way of becoming interesting, adventurous, and even dangerous."

Over the past couple of weeks, the Sassafrases had become experts in the interesting, adventurous, and dangerous. They were pretty sure they could handle the Arabian Desert.

"No problem," Tracey agreed. "What are your horses' names?"

"Ike and Wyatt," Hibbel replied.

"What time does the race start tomorrow?" Blaine asked.

"Approximately thirty minutes before sunrise. Sheikh Rehan will start the race with the wave of a flag," Sylvester answered.

Blaine and Tracey went and made their acquaintances with

Ike and Wyatt as Doc Hibbel threw out his last sales pitches to the remaining racers. After all the racers left, the twins helped Sylvester move his rolling shop and horses into the stable where the rest of the racehorses were. Then, they had dinner with the old peddler over a grill on the faux western street.

Eventually, Sylvester Hibbel headed out to his awaiting five-star hotel room, and the Sassafras twins were left alone in the stable at the edge of the desert. They both drifted off to sleep in the back of the wagon in the quiet still of the Arabian night, dreaming of horse races, performing King Crabs, and robot squirrels.

Sweating in the Desert

Neither of them knew exactly what time it was, but they both awoke with a start sometime during the night.

"What was that?" Blaine whispered, blinking his eyes. "Tracey, did you hear that?"

Tracey rubbed her eyes and peered through the dark at her brother.

"You mean did I hear that crashing noise?" she asked quietly.

"Yeah, that crashing noise," Blaine confirmed. "Like the sound of something metal hitting the ground."

"Yeah, I heard it," Tracey responded.

Both twins sat up and peeked out of the back of the covered wagon. Everything was very dark inside of the stable, but the Sassafrases could now hear the sound of an angry whispering voice.

"You fool!" the voice chided. "You dropped the bucket! We will never be able to find all the poison pellets on the ground in the dark!"

As the twins' eyes adjusted a little to the dark, they could now make out two shadows scurrying around on the ground by some of the racehorses.

"Hurry, hurry," the frantic voice urged. "Just find out what you can. A little poison will be better than no poison at all."

"Blaine!" Tracey whispered. "Is someone poisoning the horses?"

"I think so," Blaine said gravely.

"We have to figure out a way to stop them!" Tracey shout-whispered.

Blaine agreed. He summoned up his best Dad Sassafras voice and yelled out into the dark stable.

"We can see you two!"

The two shadows froze and stared in the direction of the covered wagon. Then, they darted for the stable door.

"Better hurry!" Blaine shouted in his fake deep voice again. "We are coming after you!"

The stable door opened and out went the shadows. The twins waited in silence for several minutes, making sure that whoever the two were that had been in the stable were truly gone.

"Do you think they're really gone?" Tracey asked her brother.

"Yeah, I think so," Blaine responded. "I just hope they don't come back."

The Sassafrases laid back down and tried to sleep, but they found that hard to do, considering they were still a bit shaken by the intrusion.

The signs of dawn began showing after a few hours of fitful tossing and turning. The jockeys started showing up and began pulling their horses out of the stable to get ready for the start of the race. The Sassafrases walked the stable and looked around on the ground but couldn't find any trace of poison pellets. They weren't really even sure whose horse the intruders had been trying to poison.

Sylvester Hibbel showed up looking like he'd had a fantastic night's sleep. He had the twins help him get Ike, Wyatt, and the covered wagon out of the stable. They meant to tell him about the possible poisoning, but as soon as they were outside the stable and back out onto the old western street, they got caught up in the excitement of the start of the race and completely forgot.

The jockeys and their horses all had their race numbers pinned on. Many of the horses impatiently snorted and stomped in place, looking more than ready to take off running. Some of the jockeys paced back and forth or did stretches to get loose. Some could be seen praying, presumably for favor, safety, and a good race.

"Do you two want to start the race up here with me?" Doc Hibbel asked the twins, pointing at his wagon's driving perch.

The Sassafrases were so excited they couldn't even answer.

"Okay! I'll take your bright eyes and open mouths as a 'yes'," Sylvester chuckled. "Climb on up there!"

First Tracey, then Blaine took their seats on the wooden bench. Old Doc Hibbel climbed up right behind them and then sat down in the middle of them, taking the reins.

"Blaine and Tracey Sassafras," Sylvester announced. "Are you ready for the Wind Tower 100?"

Both twins nodded slowly, full of anticipation, as they watched the horses and riders line up at the starting line. Sylvester used the reins to gently guide Ike and Wyatt to line up just behind the racers. Sheikh Rehan stepped up onto a wooden podium and slowly raised the race's starting flag.

The Sassafrases looked at the wide desert horizon in front of them. The sun looked as though it would peek up at any moment. The flag held high in Rehan's hand flapped vigorously in the wind. The jockeys crouched in position, and the horses patted their hooves on the ground. The crowd of fans held their breaths, and each person's heart raced inside of them.

Sheikh Rehan's hand came down in a flash, with the flag streaking down behind it. The race was on. The crowd cheered as the powerful horses bolted forward into a sprint. The starting line was now behind them and the Arabian desert before them.

"Heeyah!" Hibbel shouted as he slapped at the reins.

Ike and Wyatt lurched quickly forward, jerking the wagon into a cruise right behind the racers. Blaine and Tracey smiled as the desert wind whipped through their hair. Learning science was sure fun. Sylvester Hibbel also smiled, and even laughed out loud in joy. The twins knew he was having the time of his life.

Eventually, the race horses disappeared over the dunes in front of them, and Sylvester guided the stage coach off their trail toward the road he had mentioned the day before.

"The way the race is set up this year," Hibbel shared, "is that there are three checkpoints in the desert and two along the road. The Sheikh asked me to be at both roadway checkpoints with my store full of goods ready. As the riders race off for the first checkpoint in the desert, we will cut over to the second checkpoint so we can be waiting for them when they start to come in."

Almost as soon as the sun came up, the twins could tell it was going to be hot, but they hadn't really realized just how hot. Even with the wind whipping across their bodies as they bounced along sitting on the driver's bench, the heat of the sun caused them to begin sweating. Even though their skin wasn't as light as Arnie Derbinhoogan's, they were a little worried about getting sunburned.

"Well, blazin' suns!" Sylvester exclaimed as he looked at the twins. "You two sure are a-sweatin'."

"We sure are!" Tracey agreed above the noise of the lumbering wagon. "Maybe we need some of that balm stuff."

"That may be true, little darlin', but by the look of the sweat on your face, you'll need another one of my products before you'll need any balm."

"We will?" Tracey asked.

"Yep, I have all manner of products that serve to aid the Integumentary System, which is the system that covers and protects the body and excretes additional waste. Did you know that it includes skin, hair, fingernails, and sweat? Even animals like Ike and Wyatt have an integumentary system. It looks different on them, but still serves the same basic functions.

"Regarding your sweaty brows," Hibbel declared, using a handkerchief to wipe the sweat from his own brow. "The skin is covered with over three million sweat pores, or tiny holes, in its surface. Each of these pores leads to a sweat gland, which releases sweat when the body becomes too hot. This sweat flows through the pores and out onto the skin's surface, where it evaporates. The process draws heat from the body and cools it down.

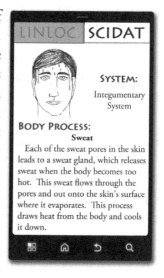

LINLOC   SCIDAT

SYSTEM:
Integumentary System

BODY PROCESS:
Sweat

Each of the sweat pores in the skin leads to a sweat gland, which releases sweat when the body becomes too hot. This sweat flows through the pores and out onto the skin's surface where it evaporates. This process draws heat from the body and cools it down.

"Although the body sweats most when it is hot, you actually sweat a little bit all the time. This is because the sweat also removes waste products from the body."

Doc Hibbel continued, "Sweat is a salty liquid made in the sweat glands. It contains water, ammonia, urea, salt, and sugar. The ammonia and urea are leftover from when the body breaks down proteins. Most of this is excreted through urine, but some is secreted through the skin. Sweat is also known as 'perspiration', and can seem to have a bad aroma. This is because the bacteria that live in your skin mix with the substances found in the sweat to produce a smell. As you get older, extra hormones that are secreted in the sweat can cause an increase in the smell. I tell you all of this to say that if you sweat too much, your body can become dehydrated from the loss of water. So it's important to drink plenty of water.

"But if you can't find water, I've got just the thing for you," the salesman pitched. "Second only to water in the rehydration game is Old Doc Hibbel's Sarsapa-Aid"

"Sarsapa-Aid?" Blaine asked. "What's that?"

"Why, I'm so glad you asked, son," said Sylvester smiling. "Sarsapa-Aid has that classic and smooth, old-timey soda pop flavor, but also has all the vitamins and minerals of any new-timey energy drink. It comes in a variety of flavors, but our three most popular are: Sarsaparilla, Root Beer, and Cream Soda. If you're a son or daughter that can't find water, try Old Doc Hibbel's Sarsapa-Aid!"

Blaine and Tracey laughed at Sylvester's funny sales jingle.

"I'll be sure to get both of you a bottle or two when we get to the checkpoint," the old peddler offered.

The twins smiled and nodded, thankful. They rolled through the sand for quite a while until they reached the roadside checkpoint, which consisted of an Arabian style tent and several watering troughs. Sheikh Rehan had already arrived in his black Hum-V limo, as had some of the other race fans who had vehicles. Sylvester Hibbel brought the covered wagon to a halt right within sight of the tent. He climbed down from the driver's box and greeted his friend the Sheikh. Then, he and the twins got the western goods and refreshments ready for the soon-to-be arriving riders.

Sylvester once again showed Blaine and Tracey how to open up the wagon's back end, so that it turned into a sales booth. He also made sure they knew where everything was and what everything was called. Just as promised, he gave them each a bottle of Sarsapa-Aid. They both thought it tasted decent, far better than the fixer elixir, but that wasn't a tall feat. After a bit of waiting, the first racers could be seen. The regal crowd cheered as the jockeys and their horses arrived at the checkpoint tent one by one.

"It is interesting to watch their different strategies," Sylvester noted. "The jockeys riding stallions tend to start fast, but finish slow, whereas the jockeys riding Arabian thoroughbreds keep a more consistent pace throughout the race. They also have different strategies at the checkpoints. The only thing that is required of them is to check-in with a race official so the official can check their race number and sign them in and out. So some of the racers check-in and then blaze through, barely even stopping to let their horse get a drink of water. Others put a little more value on rest, so they will stop at the checkpoints and stay awhile, balancing out their strength and stamina a little more. Riders with different strategies seem to win the race each year."

The Sassafrases watched all the jockeys as they came into the second checkpoint and saw that Hibbel was right. Some riders zipped through while others got off their horses and stayed awhile. A big number of those that stayed took the time to come over to the covered wagon and buy some of Old Doc Hibbel's products. Sylvester sold hats, belts, balm, and bottle after bottle of Sarsapa-Aid.

This being only the second checkpoint of five, the racers seemed to still be fairly close together, but there were a few stragglers, and the twins were sad to see that among those at the back of the pack was the European champion, Arnie Derbinhoogan. His face was downcast as he and Horsinhoogan hobbled into the checkpoint. A race official checked his number and signed him in.

Arnie then jumped off Horsinhoogan and slowly pulled him by his reins over to the covered wagon. The poor horse was shaking and was visibly foaming at the mouth.

"What's wrong?" Sylvester asked the jockey, sincerely concerned.

"I don't know," Arnie answered. "I think maybe Horsinhoogan has been... poisoned."

"Poisoned!" Sylvester exclaimed in shock.

The twins wanted to slap themselves for forgetting to tell Doc Hibbel about what they had overheard during the night in the stable.

"That may be the problem," Blaine said flatly. "We forgot to tell you that a couple people snuck into the stable last night and were talking about poison pellets."

The look on Arnie's face showed that his worst fears had come true. Sylvester still looked shocked, but only for a second. The twins could almost see the light bulb turning on in his mind as he turned and began rummaging through his products. He grabbed something that the twins recognized: a small glass bottle of Old Doc Hibbel's Fixer Elixir. Without even asking Arnie's permission, Sylvester opened Horsinhoogan's mouth and poured the entire bottle down the horse's throat.

"Now, quick!" Hibbel ordered to Derbinhoogan. "Get Horsinhoogan over to the troughs and chase that elixir down with gallons and gallons of water! If there is poison in his system, this elixir mixed with water is our best chance to expel it."

Without asking any questions, the Dutchman led his sick horse over to the water troughs. Sylvester went with him, but he instructed the twins to stay and watch the store.

"Looks like we may have a few saboteurs among our racers this year," he said gravely, before he followed Arnie to the troughs.

The Sassafrases tried to keep their minds on selling western goods to the fans and racers who were still hanging around the checkpoint, but they just couldn't help but think of poor Arnie Derbinhoogan and his poisoned horse. They kept glancing over toward the water troughs, hoping to see if the elixir was working. On one such glance, they spied something different altogether. Two men, wearing identical black robes and turbans, seemed to be exchanging their race numbers. When the twins looked closer, however, the numbers were actually exactly the same.

"Hey, they're not supposed to have the same numbers, are they?" Tracey asked, nudging her brother's arm.

"I don't think so," Blaine responded. "That sure looks like cheating to me."

The two riders parted company with one riding out into the desert in the direction of the third checkpoint. The other quietly walked his horse up onto the roadway. He waited to make sure he was going unnoticed, and then he slinked off, disappearing on the other side of a sand dune.

Before he disappeared, he made eye contact with the Sassafras twins. His hateful and angry eyes sent shivers of fear through the twins' bodies. They recognized that face! That was the face of the infamous and ruthless Itja: leader of the Kekeway!

CHAPTER 17: THE INVINCIBLE ITJA

Hair Today, Gone Tomorrow

A whirlwind of questions filled the minds of Blaine and Tracey Sassafras. How had Itja gotten to the United Arab Emirates? Had he escaped from Princess Talibah's father, Abubakar? Did he have access to invisible zip lines too? Had he poisoned Arnie Derbinhoogan's horse? The twins almost felt like falling down from the weight of all these questions. They had to tell Doc Hibbel about seeing this evil man. They both rushed over toward to the water troughs to where Hibbel and Arnie were.

"Doc! You've got to hear this!" Blaine started to say, but then he saw Horsinhoogan.

The horse, who only moments before, had been shaking and foaming at the mouth, now miraculously looked completely healthy.

"Wow! It's amazing!" Arnie marveled. "Thank you, Doc! That Fixer Elixir really does work!"

Sylvester smiled and walked up to Horsinhoogan, giving the horse an encouraging pat on the neck.

"Now you go out, catch up, and win this race," the kind peddler said to the recovered horse.

Arnie shook hands with Doc, climbed up on Horsinhoogan, turned the horse out toward the desert, and shouted, "Heeyah!" The horse and rider took off like a bullet into the thick of the Wind Tower 100.

By now, all but two or three of the racers had left the second checkpoint. They were all racing out to the third checkpoint located somewhere amongst the sand dunes.

Sheikh Rehan and most of the race fans had left, too,

traveling on down the road to catch glimpses of the racers at checkpoint number four and then at the finish line.

"Well, Blaine and Tracey, looks like it's time to pack up the wagon store and move on down the trail to checkpoint number four," Sylvester Hibbel announced.

The twins began helping the salesman put up his wares, but the Itja sighting was still plaguing their thoughts.

"We think we saw some racers cheating," Blaine told Hibbel as they folded the stairs back into the wagon.

"Cheatin'?" Sylvester asked. "What do you mean?"

"Two of the riders had the exact same number," Tracey explained. "One of them rode out toward checkpoint number three, but the other one snuck to the other side of the roadway somewhere."

Sylvester raised an eyebrow and rubbed the salt-and-pepper-colored whiskers on his face.

"Hmmm," he murmured as he was thinking. "That does sound like cheatin'. If two riders are racing with the same number, and one or even both of them is taking shortcuts, then in theory, that race number could still be officially signed in at each checkpoint. On paper, it would look like the same rider and horse completed the race, but in actuality, it would be two riders using two horses, working together to cheat their way to a victory."

Sylvester rubbed his whiskers again. "Nope, this isn't good at all. I think if we can stop them, we should. Where did you two say that cheatin' rider disappeared to?"

The twins walked up onto the roadway and found the spot where they thought they had last seen Itja. Doc Hibbel followed the twins to the spot and then inspected the area. The roadway was raised quite a bit, with mounds of sand dropping down and away from it on either side. As they slid down through the sand on the other side of the road, they looked back and couldn't even see the top of the checkpoint tent on the other side. So it definitely would have been easy enough to ride a horse on this side of the road without being seen.

"Well, I'm not a tracker, per se," Sylvester shared, after looking around for a couple of minutes, "but I'm pretty sure I have found our cheater's tracks."

The twins could see the tracks in the sand as Hibbel pointed at them.

"It looks like they follow alongside of the road in the direction of checkpoint number four," Sylvester declared. "And by the spacing of the prints, it looks like he is not running his horse, but rather taking it pretty slow. I think with the two-horse power of Ike and Wyatt, we can catch him in the wagon."

The three tromped back up over the roadway and loaded up on the covered wagon. Sylvester guided the big vehicle to where the tracks were, and then the pursuit of Itja began.

The blazing sun was even hotter now, and the whole desert seemed to be steaming around them. There was no vegetation, no wildlife—only a few scattered oilrigs and pump-jacks and mounds and mounds of white Arabian sand.

The twins could easily see why Sylvester thought the Wind Tower 100 was the hardest distance horse race in the world. They were amazed at the kind of will and stamina a racer needed to have to compete in this long race across such a harsh stretch of land. The roadway they were traveling next to, for the most part, was long and straight, but they could see ahead that it curved left toward the north.

Sylvester continued to keep his eyes on the tracks as Ike and Wyatt pulled them through the sand. When they reached the curve, the peddler-turned-tracker commented, "Well, it looks like the tracks go left around the corn—"

Sylvester's sentence was interrupted as the two normally easygoing horses, Ike and Wyatt, suddenly reared up on their hind feet and let out loud whinnies. Then, before they even really knew what was happening, the Sassafrases and Hibbel found themselves surrounded by a group of black-robed men on horses. They were shouting and swinging swords around in the air as Ike and Wyatt continued to go crazy until one of the black-robed men slowly raised his hand and signaled for the ruckus to stop.

When everything grew quiet, he prodded his horse forward and then removed the black cloth that had been covering the lower half of his face. The Sassafras twins shuddered in fear. They had been right. It was Itja.

"Well, well, well," he said in his deep, creepy voice. "If it isn't my young friends, the camel thieves."

Sylvester was clueless as to what the man was talking about, but the twins' minds immediately raced back to their first morning in Egypt, when this man had accused them of stealing his camels. In fact, he had been the true thief. How had he gotten from Egypt to the UAE? As if reading their minds, Itja addressed the issue.

"You two are probably wondering how myself and a number of my loyal Kekeway got to the UAE, are you not?" Itja asked.

The Sassafrases nodded.

"Well, I'm wondering the same thing about you two," Itja said, with his face turning from arrogant to angry. "I'm beginning to think that my enemies have sent the two of you to spy on me. Then again, two children against Itja and the Kekewey? It sounds preposterous to even say it, and yet you two keep showing up. You showed up when I nearly had myself some royal camels. You showed up when I nearly had myself a drove of fennec foxes, and now you show up here, one thousand five hundred three miles away, when I nearly have myself a weighty winner's purse. I still don't know how all of you escaped from the tomb in Egypt. Maybe you are invincible."

Itja looked at the dozen or so Kekewey bandits that were with him. "What do you think? Are these children invincible?"

None of the Kekewey members looked like they really knew how to answer their leader's question. Itja then returned his glare back to Blaine and Tracey.

"Or maybe it is I who am invincible," he said. "I kidnapped the daughter of a king, plundered his land and people, and then escaped his custody before he could put me on trial. Now I'm bold enough to cheat a rich and powerful Sheikh out of his winner's purse. Yes, maybe it is I who am invincible!"

Itja made a hand motion to his men, and they immediately moved in on the three. They pulled them off the wagon's driver's perch, and then they started tying them up.

"Lucky for us today," Itja sneered with the arrogance back on his face, "we have a way to put this invincibility to the test."

After Blaine, Tracey, and Sylvester were successfully tied up, they were led by Itja and the Kekewey to a nearby see-sawing pump-jack. As the twins watched the metal arm of the machine go up and down, they wondered why they'd been brought to this strange location.

"Here is your test," Itja declared. "Your test of invincibility!" Blaine and Tracey had no idea what the evil bandit was talking about, and by the look on Hibbel's face, he didn't either.

"But before I explain to you how your invincibility test works, let me explain to you mine." Itja pulled out the cloth race number that the twins had seen earlier, and then he called one of the Kekewey over. He pinned the number to the man's black robe and then shouted at the man, "Ride!"

The Kekewey bandit kicked his horse and then took off back toward the roadway.

"My invincibility test is this," Itja said to his three captives. "I must scheme well enough to win this race, the Wind Tower 100. Needless to say, this is a very difficult test. However, with multiple horses, identical looking riders, and two copies of the same race number, I am confident that I have a fail-proof plan."

Itja then turned everyone's attention back toward the pump-jack.

"Your invincibility test is a little different." The man laughed wickedly. "It includes this pump-jack that you know see in front of you. Some of my men have, shall we say, 'customized' it specifically for the three of you. You each will be tied to the ground, next to the spot where the bridle goes in and out of the ground attempting to pump up oil and gas. This will put you directly under the big heavy end piece of the rocking metal beam called the 'horse head'. In the beginning, it will be no threat to you as it teeters back and forth. But every time that it pumps away from you, its beam will click down a notch. So, when the horse head comes back in your direction it will get just a little closer to hitting you. Eventually, the customized pump-jack will click down enough notches and the three of you will be smashed where you lie."

"How is that a test?" Sylvester asked, speaking up. "Sounds more like torture to me."

"Call it what you may, but if you can escape this specially designed smashing machine, and stop me from winning the race, then you will prove yourselves to be truly invincible," the bandit bully taunted. "But might I just say that your chances of passing your invincibility test seem a little slimmer than mine."

The members of the Kekewey that had made it from Egypt to the UAE with Itja laughed along with their scoundrel leader as they dragged the three captives over to the pump-jack and tied them securely down to the ground. Itja gave the Sassafrases and Hibbel one last arrogant stare, and then without saying another word, he and his men rode off into the desert in pursuit of their dark-hearted dream.

The twins lay there, fastened to the hot desert ground, in complete silence. They felt scared, and they felt responsible for getting Old Doc Hibbel into this mess. They should have known not to mess with Itja again.

The metal beam of the pump-jack rocked back and forth, back and forth. The twins heard the sound of a click as the beam clicked down a notch. When the horse head came lunging back their way, it was just a little bit closer to hitting them. Blaine and Tracey lay there, helpless and gulping, and every time they heard the dreaded click, they couldn't help but jump.

Sylvester Hibbel, however, started chuckling. "Well, I'll be a broken string on Gene Autry's guitar. This is quite a conundrum we've gotten ourselves into, isn't it?"

The twins didn't know why—maybe it was Doc's good attitude, maybe it was a nervous reaction—but they started laughing, too.

"Yes, sir, it sure is," Blaine agreed.

"Yep," Tracey added. "This is a regular, smashing good

time."

All three started laughing harder at Tracey's pun. They figured it was better than crying.

Time passed. They told some more jokes, they exhausted escape ideas, and the evil oil machine continued to click farther and farther down. Now, when the horse head came swinging down their way, it was only inches from touching their bodies. They all figured they only had a couple more notches left.

"Sure wish I had some elixir that could remedy this situation," Sylvester vocalized, now sounding sad.

Just then, a bald, white-robed Arab man came hobbling up to them, pulling his horse behind him with one hand and brandishing a dagger in the other.

"My friends!" he exclaimed anxiously. "Let me help you!"

He dropped to his knees and used his dagger to quickly cut away the ties that bound the three to the ground. The twins and the salesman hurriedly rolled out from under the plunging horse head. It came down toward where they had been lying, and then went back up, as the machine clicked into its final notch.

This time, when the horse head came back toward where they had been tied down, it hammered down faster and smashed with force right into the ground, creating a small crater. The metal beam then broke from its perch, falling off the rig, and landing in a dusty sandy cloud on the ground. Blaine and Tracey sat and stared at the broken pump-jack.

"Thank you so much, my kind, bald friend," Sylvester expressed to their rescuer. "We would have been lunchmeat if you hadn't come along."

The man dipped his head, acknowledging the gratitude.

"How did you find us out here?"

"I was the last rider to check-in at checkpoint number two," the man shared. "When I arrived, I asked the officials where your wagon store was as I needed to buy something. They pointed to the other side of the road. I followed your tracks, found the wagon by itself, and then found you here."

The three looked and saw that the covered wagon was still visible off in the distance. Surprisingly enough, Itja and the Kekewey had left it alone.

"Well, if there is anything at all that you need from my store, I will be more than happy to accommodate you. And free of charge, of course! That's the least I could do for you after you saved our skin. What is it that ails you, my good man?" Sylvester exclaimed.

The white-robed racer's face was covered in a sudden sadness.

"It's my horse," he shared. "I think he's been poisoned."

The three looked closely at the man's horse for the first time. They couldn't believe they hadn't already noticed it because the horse was showing the same symptoms of poisoning that Horsinhoogan had shown earlier. Shaking and foaming at the mouth.

"Oh, no!" the Sassafrases thought. "This horse must have eaten some of those poisonous pellets, too."

"Is there anything that can help him?" the man asked desperately.

"There sure is!" Sylvester responded.

The peddler then took off in a blistering sprint back toward the wagon. The twins still didn't know how old Sylvester Hibbel was, but the Oklahoman sure moved like he was younger than he looked. He reached the covered wagon and then drove it back to where they were.

He hopped off, reached in the trunk, pulled out a bottle of fixer elixir, and ran over to the racer and his sick horse. He poured the medicine down the animal's throat as he instructed Blaine and Tracey to go grab as many bottles of Sarsapa-Aid as they could find.

The twins rummaged through the back of the wagon following Sylvester's instructions.

"If you're a son or a daughter that can't find water, try Old Doc Hibbel's Sarsapa-Aid," Blaine recalled as they ran back with arms full of the drink.

"Works for horses, too," Tracey added with a smile.

Sylvester grabbed the bottles from the twins two at a time, popped the tops off, and then poured them into the horse's mouth.

"What's your name, my friend?" Hibbel asked the jockey as they all waited to see if the elixir would work.

"My name is Najib," the man answered. "And my horse's name is Yazer."

They all watched as Yazer slowly stopped shaking and his foaming at the mouth ceased. Within a matter of minutes, he looked good as new. Yazer lifted up on his back two legs and let out a powerful neigh, as if saying he was ready to race again. A huge smile found its way to Najib's face.

"Yazer boy!" he exclaimed. "You're all right!"

The racer hopped up on his steed.

"Well, then," he said to the three. "I guess we'll finish the Wind Tower 100 after all."

The twins and the peddler smiled.

"Is there anything else I can give you?" Sylvester asked. "Is there anything else you need? Anything at all?"

Najib thought for a moment, and then reached up and touched his bald head. "The wind did blow my head scarf off at the beginning of the race. I heard some of the other jockeys talking about something you have called 'Old Doc Hibbel's Hide-a-Chap Western Balm'. Do you think that would help my bald head from getting sunburned?"

Sylvester's eyes lit up. "I've got something even better," he responded.

He dug through the wagon's trunk until he found what he was looking for.

"Old Doc Hibble's Hair-be-Here Sproutin' Sap," he said, holding up another one of his old-timey-looking glass bottles. "Just pour this solution on the desired area, and after an hour or two, hair be here, hair be there, hair be everywhere! This toutin' sproutin' solution has been known to put hair on men and women in different stages of balding. It'll sprout hair on a shiny head, it'll sprout hair on a thinning head; shoot, it'll even sprout hair on a watermelon if you want it to. Old Doc Hibbel's Hair-be-Here Sproutin' Sap!"

Sylvester threw the bottle up to Najib, who caught it in the air.

"Thanks, Doc," he smiled.

"Now, get out there and finish the race, friend!" Doc Hibbel ordered.

Najib nodded as he prodded the rejuvenated Yazer. They rode off to finish the race. Tracey watched the horse and rider ride off and wondered if they would be able to finish the race in the set time period. On the other hand, Blaine was wondering if that sproutin' sap stuff would give him a mustache.

Once they were back in the covered wagon and heading in the direction of checkpoint number four, Blaine had a question for Doc Hibbel, "Does that hair sap really work?"

"Well, before I tell you if Old Doc Hibbel's Hair-be-Here Sproutin' Sap really works, let me tell you how hair, in general, works."

"The body is covered with hair, except the lips, palms of the hand, and soles of the feet. Some places on the body have more hair follicles than others," Hibbel explained. "For example, the head has over one hundred thousand follicles. Some hair is easily visible, like the hair on your head and eyebrows, but some is so fine that it is difficult to see, like the hair on your ears. Either way, all hair grows from living cells that are at the base of the follicle. These follicles are found in the layer of fat at the bottom of the skin. As the cells push upward from the base, they die and are filled with a tough substance called keratin. The part of the hair that we see is dead, which is why it doesn't hurt when we get it cut."

"The average person loses fifty to one hundred hairs every day, but these hairs are usually replaced with new ones that grow from the same follicle. However, as we grow older, not all of these hairs will be replaced. At the bottom of each hair follicle you'll find a sebaceous gland, which secretes a natural oil called sebum. This oil coats the hair to make it shiny and waterproof. It also helps to keep your skin soft and flexible. I have one last thing to say about hair, but first—have you ever had goose bumps?" Doc Hibbel asked the two twelve-year-olds.

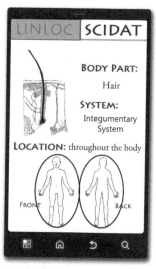

"I'm pretty sure I got some when I found out we were going to be tied down under that horse head thing," Tracey confirmed, as the two both nodded.

"I think that I did too," Hibbel responded. "Well, you will sometimes see goose bumps on your skin when you are scared or cold. These appear because tiny muscles in your skin pull the hair follicles so that they stand upright, causing a bump to appear on the skin. This helps the body trap warm air in between the raised hair and the skin."

Sylvester reached up and briefly stroked his mustache as he went on, "So now, back to your original question, Blaine—Does Old Doc Hibbel's Hair-be-Here Sproutin' Sap really work?'"

Sylvester paused and looked at Blaine who was sitting next to him on the driver's perch. "The truth is I don't know. I have never sold a bottle of it nor have I ever given it away, so I just don't know."

Blaine sighed and rubbed his upper lip. Maybe someday he would have a mustache.

Fingernail Boo-boo

They made it to checkpoint number four, but it was much later than Sylvester had wanted to. The whole getting ambushed and tied down next to a jerry-rigged pump-jack thing had put them more than just a little behind schedule. The peddler was sure that a large number of the racers had already checked in here and left, on their way to checkpoint number five. There were still a handful of race fans with their vehicles here, but Sheikh Rehan was nowhere to be seen. He was probably in his Hum-V limousine right now heading to Hatta and the finish line.

After the three parked the covered wagon, they headed straight over to the race official's table inside the checkpoint tent to report Itja and his gang of cheaters. The residing race official listened to their explanation of how Itja was using two copies of the same race number as well as more than one horse and rider to illegally check-in and shorten his race times. The official listened intently until they were done and then got a curious look on her face.

"That's strange that you three are reporting this," she declared. "Only about thirty minutes ago, another rider came in and reported other racers trying to cheat using these same tactics. He actually gave me fifteen specific race numbers that he said had been copied and were being used by teams of racers working together."

The three left the tent shocked. Not because of widespread cheating but because they knew exactly who the man was that had reported the alleged cheating.

"Wow, that's a pretty tricky move," Sylvester whistled. "By

lying and reporting fifteen cheating race numbers, this Itja character is hoping to create confusion at the finish line, enabling him to slip in first amongst the chaos and claim the winner's purse."

"We've got to stop that from happening!" exclaimed Blaine.

"I agree," nodded Hibbel. "Let's make sure that the western goods are available for some of the slower racers who have yet to check-in. Then, let's high-tail it to the finish line to try and prevent this black robed bandit from cheating his way to a victory."

This sounded like a good plan to the Sassafrases, so they headed back over to the rolling store and did what they could do to help the Doc sell his goods.

They saw Arnie Derbinhoogan ride up fast and check-in. Thankfully, he looked good and Horsinhoogan looked healthy. He smiled and waved to the three as he led Horsinhoogan over to the troughs to get a few long gulps of water. Wasting no time, he then took off back out into the sand toward the last checkpoint.

"By executing such a short check-in, he probably just passed about a dozen racers," Sylvester informed. "I know that in European circles Arnie is known for lightning fast finishes. It looks like he may have a chance to win this race yet."

The twins stood there at the back of the wagon with the salesman a while longer and watched him continue to give away about as many products as he sold. Eventually, he made the motion that it was time to close up shop and head for the finish line. Just as he did, the biggest, strongest, most muscle-bound Arab man the twins had ever seen came slowly riding up on his horse toward the wagon.

He had a stern scowl on his face, and he looked tense and mean. The twins gulped, immediately intimidated. He stepped off his horse, his sandaled feet landing firmly on the sand. He was so big that even while standing on the ground, he was much taller than his animal. He wore a dark brown robe and headscarf and had a huge sharp sword hanging at his side. His muscles flexed in the sun as he glared at the three.

Then, faster than one would think possible, the big man... burst into tears.

"I cracked my fingerprint!" he cried out through his sobs.

"Cracked your fingerprint?" questioned Sylvester. "That doesn't make any sense. Your fingerprint is the skin pattern on the tips of your fingers that is unique to you. The skin on your fingers, toes, palms of your hands, and soles of your feet is folded into tiny ridges, making swirling patterns. These ridges help the hands and feet grip. I'm sorry, my big weeping friend, but it's just not possible for you to have cracked your fingerprint."

"But look," the man whined, holding up his hand for all to see. "Look at my boo-boo!"

As the crying man held his hand up, knuckle side facing them, the twins could now clearly see that the fingernail on his ring finger was split right down the middle. They cringed. It looked painful, but probably not painful enough to cry about—especially for a grown man.

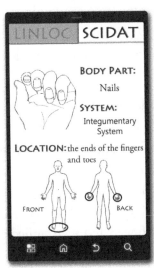

"Ohhh," Sylvester acknowledged. "You cracked your fingernail."

The man nodded, big tears running down his cheeks.

"The fingernails protect the fingertips, which are sensitive because they have a large concentration of sensory receptors, which is probably why this cracked fingernail is hurting you so much, big fella. Fingernails also help us pick up items or scratch an itch we might have. Our nails are made from keratin and they grow from the nail root, which is underneath the cuticle at the base of the fingernail. The cuticle helps to protect the new nail as it grows out. Over time, the new cells harden into older cells that form the fingernail. The older cells remain attached to the nail bed until they reach the end of the finger. This nail bed is fed with tiny blood cells, which gives the nail its pinkish color."

"The typical fingernail grows about 0.12 inches per month, but they can grow faster in the summer than they do in the winter. It takes about three to six months to completely replace a nail, and if you don't trim your fingernails, they can grow up to a yard in length, curling as they grow out."

Blaine and Tracey made squeamish looking faces at the thought of long curled fingernails.

"The body also has nails covering the toes," Sylvester shared, "which grow three to four times slower than those on the fingers."

"So what does our friend need for his hurt fingernail?" Tracey asked. "Do you have some kind of balm or elixir or sproutin-sap that he can put on his finger to help heal it?"

"What this man needs..." Hibbel answered while digging through is wares, "...is..."

The twins listened intently as the salesman spoke, interested to hear about his next off-the-wall but effective old western product.

"...a band-aid."

"A band-aid?" the Sassafrases exclaimed together. That is not what they expected the old peddler to pull out.

"Yep, a plain old band-aid," Sylvester affirmed. "It'll help prevent this big guy's finger from getting infected."

Hibbel took the packaging off of the band-aid and wrapped it securely around the big rider's hurt fingernail. The big man stopped sniffling, looked at his bandaged finger, and then smiled at Sylvester.

"Thanks, Doc," he said.

Then he got on his horse and rode off.

"You shouldn't bite your fingernails, either," Hibbel added, still thinking about the science. "You can open up the skin and make it vulnerable to infection as well. Plus, your fingernails collect a lot of germs, which can enter your body through your mouth when you bite your nails."

The twins nodded in understanding before asking, "Now, off to the finish line?"

"Off to the finish line!" Sylvester said, as they climbed back up into the driver's perch of the covered wagon. Hibbel slapped the reins against the backs of Ike and Wyatt, and onward they went as the sun began to set in the west behind them.

As they rode on, the Sassafrases examined their own fingernails and used their phones to take pictures. While they were at it, they also took some pictures of their hair using their new microscope app.

"The last two legs of this race may be the hardest for the jockeys and their horses," Sylvester explained, as they rolled down the road. "Checkpoints one through four are out in the desert. After checkpoint number four, the trail rises fairly quickly up into the Hajar Mountains. So, checkpoint number five is in the mountains and then from there to the finish line, the trail, with the exception of the last three hundred yards, stays entirely in the mountains. At this point in the race, the riders aren't dealing with the hot bright sun anymore, but they have to deal with a zigzagging rocky mountain trail in the dark. This is just another one of the aspects of the Wind Tower 100 that make it the greatest and most difficult horse race in the world."

The twins smiled from their bouncy perch at the front of the covered wagon. They were both loving the experience of being a part of the Wind Tower 100. So much, in fact, that they wanted to stay and see the conclusion of the race, even though they were finished gathering their SCIDAT data.

It was dusk, but the Sassafrases could clearly see the craggy Hajar Mountains that Hibbel had just talked about laying across the horizon in front of them. They were ominous-looking mountains with sharp, pointy, wind-battling peaks rising into the sky. The road they were traversing took them up into the mountains just about the time the sun disappeared completely, shrouding the desert in darkness.

The twins, who just moments prior had been enjoying being part of this race, now wondered if maybe they shouldn't go ahead and text in their SCIDAT data so they could zip out of this scary place. The Arabian wind zipped over and through the different surfaces of the mountains, making haunting and mysterious noises. Blaine and Tracey spied many cave-like crevices in the rock as they rolled on. They wondered if Itja had stationed some of the Kekewey here to ambush them again.

Around every bend of the nighttime mountain road, the twelve-year-olds expected something to happen, but no such predicament played out. Sylvester Hibbel guided Ike and Wyatt safely and swiftly through the mountains to Hatta.

When they reached the finish line, they saw that none of the racers had arrived yet. The crowd of fans was sitting on a half circle of stone-hewn bleachers, expectantly waiting for the first horses and riders to show up. Huge spotlights had been set up to illuminate the night at the finish line. The twins could see where the horse trail came out from between two rocky walls and then sprawled out onto the last three hundred yards which was on flat ground.

Sheikh Rehan sat on a raised stone seat, just under a huge wind tower that was built on a peak right here on the mountaintop. It looked similar to the wind tower they had seen built on top of the mall in Dubai but was much more rugged.

The Sheikh raised his hand in greeting and smiled to the

three as Sylvester brought the wagon gliding to a halt in a designated parking area for him.

"I was debating in my mind," Hibbel told the Sassafrases, "whether or not to go up now and tell the Sheikh about this Itja character and his cheating scheme. Now that we are here, I think we should just wait and see how this whole thing plays out. Who knows? Itja's plan might not work. Maybe Arnie or one of the other jockeys can beat him in spite of his devious plot."

The twins hoped so. They had no desire to see Itja celebrate as a victor, even for a second.

"I'm just glad," Sylvester added. "that in spite of our delays, we made it in time to see the end of the race."

The peddler pulled a pocket watch out of the front pocket of his vest and gave it a gander. "The race's leaders should be arriving any minute now."

The Sassafras twins joined the crowd of race fans in their collective hushed anticipation. All eyes were now on that narrow opening in the rock where the racers would be appearing.

CHAPTER 18: BACK TO UNCLE CECIL'S

The End of the Race

The sound of horse's hooves echoed up and out of the narrow passageway. The first racer was quickly approaching. The twins' hearts jolted with excitement when Arnie Derbinhoogan exploded out into the light. The crowd gasped in excitement. "Wow!" exclaimed Blaine and Tracey. Arnie had done it. He had come from the back of the pack to first place. It truly was amazing.

However, another horse and rider burst out into the light, right behind Arnie and Horsinhoogan. This second place rider was none other than the black-robed Itja himself. He held a determined scowl on his face and a brown leather bullwhip in his hand, which he was using to try and pop Arnie and Horsinhoogan.

"Hey," Tracey shouted. "Can he even do that? That's not allowed in the race, is it?"

The crowd began to become animated along with Tracey. They booed the dirty black-robed cheater. Arnie was successfully dodging the snaps of the bullwhip and remained in the lead.

Then, something that no one expected happened: a third horse and rider came shooting out into the spotlights like a bullet. This third duet was going noticeably faster than both Arnie and Itja. The horse's head pumped up and down as it hit the last three hundred yards of the race in a sprint that was so fast it was almost unbelievable. The jockey looked oddly familiar to the twins.

"It's Najib!" shouted Blaine.

"It sure is," Sylvester exclaimed. "But look—now he has a flowing head of hair!"

Tracey looked and saw that both Blaine and Sylvester were

right. Evidently, Old Doc Hibbel's Hair-be-Here Sproutin' Sap really did work. Najib's long flowing locks bounced gracefully in the wind as he and Yazer began to close the distance between themselves and the two leaders.

Itja saw Najib catching up to him out of the corner of his eye. He turned the attention of his bullwhip from Arnie to this new threat to his victory. However, in Itja's first attempt to pop the long-haired racer, Najib grabbed the end of the bullwhip with his bare hand and gave it a hard yank, pulling Itja right off his horse.

The scoundrel leader of the Kekewey tumbled off of his horse and rolled head over heels to a crashing stop in the sand. Even with the scheming and cheating, Itja still hadn't managed to win the race. The twins wondered how invincible the bandit felt right now.

Najib looked straight ahead and focused on catching the flying Dutchman and his speeding horse. There was now less than one hundred yards left to go in the Wind Tower 100, and, by the looks of it, it was going to be a photo finish. Horsinhoogan snorted and grunted in gritty determination. The muscles in Yazer's neck flexed over and over as he pumped his head even harder for the finish.

Najib's hair glistened beautifully under the shining of the spotlight. Arnie's un-sunburnt brow glistened with beads of sweat that were seeping out from under his Hide-a-Chap Western Balm. The finish line was now just feet away.

Everyone in the crowd held their breath. With one final, resolute push, Yazer leaned his head forward and lunged across the finish line. First. The crowd erupted in applause.

The best place the twins had found to get away so that,

when the time came, no one would see them zip away, had been inside Sylvester Hibbel's old western covered wagon. That is where the Sassafrases were now, finishing up their SCIDAT entry. This leg of their summer adventures had been amazing. They had learned all about the Integumentary System and been able to be a part of a truly awesome distance horse race.

To everyone's shock and surprise, the previously unknown Najib and his determined horse, Yazer, had won this year's Wind Tower 100. The gracious second place racer, Arnie Derbinhoogan had been the first to congratulate the victorious Arab. He even got to hand him the winner's purse.

Also, to everyone's shock and surprise, Sheikh Rehan had the cheating Itja and his Kekewey bandits arrested right there at the finish line, in front of everyone. Blaine and Tracey just hoped the Sheikh would have better luck than Abubakar, Princess Talibah's father, at actually keeping these black-robed scoundrels behind bars.

Doc Hibbel had enjoyed another good chance at the finish line to sell his goods, but the twins knew by now that the Oklahoman's real joy was just being a part of the race. They understood that feeling because they felt the same way.

"OK, I'm finished entering in the data," Blaine stated. "Let's add in our pictures and press send."

"Done," Tracey announced, as she pushed the tab on her phone that would send the pictures to the data screen down in Uncle Cecil's basement.

"I wonder where we're going next," the girl asked aloud, as she flipped over to her LINLOC app.

"No way!" her brother exclaimed, beating her to the answer

as he looked at the same app on his phone. "We're finished with anatomy! LINLOC says that Uncle Cecil's basement is the next location!"

"You're right!" Tracey confirmed, now seeing the same thing on her phone. "There is no local expert listed, or any anatomy topics to study. We're done! We did it! We successfully completed another subject!"

The Sassafras twins joyfully put on their zip lining gear and calibrated their carabiners for home. They zipped with light and smiles on their faces, disappearing from inside of the wagon, appearing seconds later in Uncle Cecil's basement.

"Train and Blaisey!" They heard their Uncle's voice say before they could stand or see. "You did it again! You successfully completed another area of science. You two are simply super-tab-derful! I'm so impressed and proud of you!"

As soon as the twins' senses returned to normal, they rushed over and gave their crazy uncle a hug. They were about to tell him that he was super-tab-derful too, when something very strange caught their eyes.

"Whoa! Uncle Cecil! What's up with the bedsheet-wearing mannequins?" Blaine asked in surprise.

The twins knew that the basement of their uncle was always cluttered, and they always expected to see strange things lying around down here, but these two sheet-wearing mannequins stood out like sore thumbs.

"Let me introduce you, or should I say reintroduce you, to Socrates and Aristotle!" Cecil said jubilantly.

"Socrates and Aristotle?" Tracey asked perplexed. "I thought Socrates and Aristotle were those two plastic skeletons of yours?"

"They are!" Cecil smiled like he was presenting a surprise gift. "These two mannequins are them! And yes, those are bed sheets, but they are supposed to look like togas."

Uncle Cecil brought the twins over for a closer look at the plastic statues.

"Before the two of you zipped away to start studying anatomy, I pulled two skeletons out of my closet. These are those two skeletons! Isn't it wonder-azing?"

The twelve-year-olds nodded.

"What the two of you didn't know was that, as you were zipping through anatomy and sending in your data and pictures, President Lincoln and I were simultaneously adding plastic anatomy pieces to the skeletal frames of both Socrates and Aristotle. For instance, when you sent in your SCIDAT data on the respiratory system, we added plastic lungs to the skeletons. When you sent in your SCIDAT data on the digestive system, we added yards and yards of spongy plastic intestines to the skeletons."

He went on, "We continued adding plastic pieces every time you sent in data until we eventually had two plastic models that looked very similar to real *Homo sapiens*, both inside and out. Whenever the topic that you were studying couldn't be easily represented by a plastic piece, we just slapped a sticker on these two guys in the appropriate place to let us know that you had covered that too. For example, if you open up Socrates' mouth and look on his tongue, you will see a sticker that says 'taste', one of the five senses you two studied in Venice."

Full of curiosity, Tracey did what her uncle was just talking about and opened the mannequin named Socrates's mouth. Sure enough, right there on his tongue was a sticker that said 'taste' on it.

"This is really cool, Uncle Cecil!" the girl exclaimed.

Both twins looked over the detail of the two mannequins. It was neat to see all the anatomy they'd studied be fleshed out for them.

Bonus Data

"This was our new surprise for you," Cecil told them happily. "But don't forget our old surprise. Why don't you two pull out your smartphones and check out the bonus data you received for successfully completing this anatomy leg of your summertime zip lining adventures!"

"Oh, yeah," Blaine exclaimed. "The bonus data! I forgot we get bonus data when we make it back to the basement."

Both twins unzipped their packs and pulled out their phones. Sure enough, they had both received texts entitled, "Bonus Data." Blaine began reading aloud.

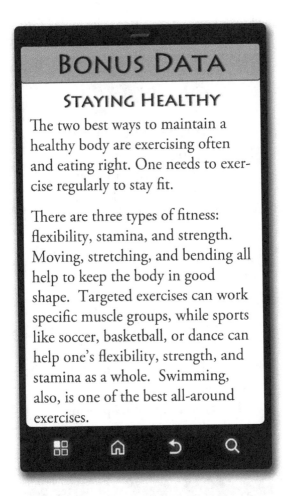

BONUS DATA

STAYING HEALTHY

The two best ways to maintain a healthy body are exercising often and eating right. One needs to exercise regularly to stay fit.

There are three types of fitness: flexibility, stamina, and strength. Moving, stretching, and bending all help to keep the body in good shape. Targeted exercises can work specific muscle groups, while sports like soccer, basketball, or dance can help one's flexibility, strength, and stamina as a whole. Swimming, also, is one of the best all-around exercises.

Blaine paused and considered the bonus data he read.

Uncle Cecil added, "So true, Blaine, exercise is important to staying healthy. Doing this correctly means that we have a balanced but consistent approach to the three key areas of fitness. But there's more, so don't stop there."

Blaine looked over at his sister as he said, "Your turn, Trace."

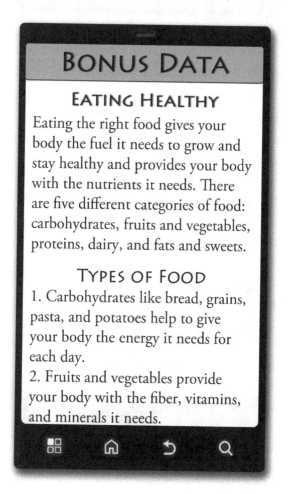

BONUS DATA

EATING HEALTHY

Eating the right food gives your body the fuel it needs to grow and stay healthy and provides your body with the nutrients it needs. There are five different categories of food: carbohydrates, fruits and vegetables, proteins, dairy, and fats and sweets.

TYPES OF FOOD

1. Carbohydrates like bread, grains, pasta, and potatoes help to give your body the energy it needs for each day.
2. Fruits and vegetables provide your body with the fiber, vitamins, and minerals it needs.

Tracey paused as she scrolled down through the bonus data on her phone. Once she found her place, she continued reading.

BONUS DATA

TYPES OF FOOD

3. Proteins like beans, fish, chicken, and other meats help your body to grow and repair itself.

4. Dairy helps give your body strong bones and teeth.

5. Fats and sweets, which your body needs but only in small amounts.

Eating right means having a balanced diet that contains the correct proportions of these different types of food.

Cecil gave a big cheer and a clap as his niece finished reading the bonus data.

"Awe-surrific!" the crazy, red-haired scientist praised. "It's great to read about the importance of exercise and healthy eating in your anatomy bonus data, but Linc-Dog and I wanted to actually give you a chance to put into practice what you just read about."

The scientist pointed to the prairie dog. "The Prez has already prepared a healthy and balanced lunch for you guys up in the kitchen. You two probably remember how he converted the kitchen into a breakfast-making machine. Well, it turns out it can make more than just breakfast. It's also capable of preparing lunch, dinner, and scrum-diddly-umpcious in-between snacks."

Cecil paused to actually rub his tummy and smack his lips, "However, before we enjoy the bounty that our furry friend has prepared for us, I have another surprise to show you! Follow me!"

Cecil ran up the basement stairs and entered the house. The twins did all they could to stay right behind their uncle as he was excited and moving fast. It was amazing that a grown man could move so quickly even while wearing bunny house slippers. Cecil led them through the ground floor of his messy house until they reached the living room. He knelt down and put some kind of big black rectangular thing inside of a squeaking machine.

"What's that?" Tracey asked.

"It's a VCR," Cecil replied.

"A VCR?" Blaine questioned. "Never heard of it."

Cecil stood up and turned back toward the twins and smiled. "It's how we used to watch movies."

The elder Sassafras turned on an old dusty television set that was sitting on top of the machine he'd called a VCR. "Your next surprise," their uncle shared with a goofy grin and big waves of his arms, "is a wonderful way for you two to practice physical fitness. Behold, I present to you, Sasser-cise!"

"It's an exercise video featuring yours truly and President Lincoln doing funky cool dance moves to the best jazz tracks of the sixties, seventies, and eighties!" Cecil said exuberantly.

Blaine and Tracey looked at each other with smirks on their faces. Just when they had started to think their uncle was fairly cool, he'd busted out his homemade exercise video. When the old TV set finally became fully illuminated, the twins saw with their own eyes what their crazy uncle had just described. The red-headed scientist dressed in a lab coat and bunny slippers and his sidekick prairie dog were dancing for exercise.

"C'mon, Train! Let's go, Blaisey!" Cecil shouted joyfully as he watched the video and started following its dance moves. "Join in with me!"

The twins just stood there, somewhat shocked, watching their uncle dance. Could they really join in with all this silliness? President Lincoln suddenly showed up in the living room, popping out from one of his holes in the wall. The prairie dog jumped right in and joined Cecil in Sasser-cising.

Blaine looked at Tracey and shrugged.

"Well, if the prairie dog can do it..." he started.

"...then we can do it too!" Tracey finished her brother's thought.

The Sassafras twins started slowly, trying to watch the video and do the correct dance moves at the correct times, but it didn't take them long to catch on. Soon they were Sasser-cising like old pros and loving it.

After about half an hour of dancing to dated jazz music, a sweaty but happy Cecil clicked the television set off and motioned for the twins to follow him into the kitchen. "We have exercised right, now let's go eat right!"

The twins followed, and soon they were sitting at the kitchen table eating a delicious, balanced, and healthy lunch prepared by President Lincoln and his amazing meal-making machine. As they sat enjoying lunch and enjoying each other's company, Blaine and Tracey got the chance to tell their uncle all about the adventures they'd had while studying anatomy.

They told him the details about everything and everywhere from Addis Ababa to Hatta. Blaine and Tracey could tell their uncle sincerely enjoyed hearing about it all.

He couldn't remember the last time he'd smiled this big. Right now, as he sat in front of a monitor down in his basement at 1108 N. Pecan Street, he was experiencing something similar to happiness. Not happiness because of anything good that had happened, but happiness because of something he hoped would happen. He was now beyond sure that he had his best plan for vengeance to date.

Looking through the lens of one of his speaker-equipped hidden cameras, he had just watched and heard those twins tell Cecil Sassafras about their adventures while studying anatomy, using loads and loads of detail. So much detail, in fact, that they had unknowingly revealed a valuable secret to his hidden listening ears.

"So, that's how that thing works," he chuckled with a devious smile as he thought back on his attempts to stop those twins while in Sydney, Australia.

He stood up from his spot at the computer and reached over and grabbed his phone. He had recorded all the coordinates for the different line locations and now he used his finger to scroll through them.

"There it is," he declared as he found the numbers he was looking for.

He jumped over to where his harness and carabiner were. His mind raced through some of his various attempts at sabotaging those twins as he slid his harness on. Leaving them marooned amongst wild animals in Kenya...trapping them in a tomb in Egypt...chasing them with robot squirrels in Alaska...none of his ideas had worked up to this point. Nothing he'd tried had managed to stop them.

He was confident, though, that this idea would let him reign victorious in revenge. Now that he knew how the thing worked, he was confident he could stop them. And when he stopped those kids, his vision of crushing the dreams of Cecil Sassafras would be complete.

Strangely enough, after recapping all of their most recent adventures with their uncle, the twins felt invigorated, not tired. One would think that after zipping to eight different locations all over the globe and getting caught up in all kinds of perilous situations, two kids would be tired, but they weren't. Blaine and Tracey Sassafras were ready for more.

They were ready for more adventure, and for more science. When they had asked their uncle what was next, he had smiled

THE SASSAFRAS SCIENCE ADVENTURES

and ran outside into the backyard. They now found themselves full of curiosity, chasing him out the door. When they made it outside, they found their uncle underneath a big, healthy maple tree, swinging on a horse-shaped tire swing.

"I-yip-I-yo-ee-yay!" Cecil shouted merrily, the wind whipping at his hair and lab coat as he swung back and forth.

"This is what's next?" Blaine asked. "We're studying tire swings?"

Cecil laughed and shook his head. He then pointed up into the branches of the big tree.

"This is what's next!" the scientist responded, pointing to the tree. "Botany, the study of plants!"

Phil Earp walked excitedly back to his dressing room. This was his first performance since getting close to winning the 'Take Our Breath Away' talent competition. Though he hadn't won the competition at the Opera House, he'd been greatly encouraged by his fellow competitors. Their kindness had spurred him to continue on as the amazing disappearing magician known to the world as the 'Dark Cape.'

Once inside his dressing room, Phil walked directly over to the wardrobe to retrieve he special suit. He pulled the wardrobe's doors open and reached in to grab the 'Dark Cape' suit, but his hand grasped nothing but air. He opened the doors wide in alarm and looked inside the wardrobe. It was empty!

There was absolutely nothing here. His suit was missing.... again!

STAY IN TOUCH WITH THE SASSAFRAS TWINS!

The adventure doesn't have to end just because you've finished the book! Connect with the twins and the other characters of the series through the Sassafras Science blog. You'll find articles in which:

* ★ Uncle Cecil explains how to extract DNA;
* ★ The Prez shares his wrap-up videos;
* ★ Doc Hibbel details the science of fingerprints;
* ★ Burly Scav tells all who will listen why your stomach growls.

Plus, Blaine and Tracey regularly pop in to say hi and share their thoughts. The Sassafras Science blog is *the place* to get to know the characters of the series!

The Sassafras twins would also love to keep in touch with you through their Facebook page. They share updates about future books, fun science-related activities, and cool nature news!

VISIT SASSAFRASSCIENCE.COM AND CLICK ON THE "BLOG" TAB TO DISCOVER MORE!

THE SASSAFRAS SCIENCE ADVENTURES